# Empowerment and Social Justice in the Wake of Disasters

T0353559

This book taps into discussions about social vulnerability, empowerment, and resistance in relation to disaster relief and recovery. It disentangles tensions and dilemmas within post-disaster empowerment, through a rich ethnographic narrative of the work of Occupy Sandy in Rockaway, New York City, after Hurricane Sandy in 2012. It details both a remarkable collaborative relief phase, in which marginalized communities were empowered to take active part, as well as a phase of conflict and resistance that came about as relief turned to long-term recovery.

This volume particularly aims to understand how community empowerment processes can breach pre-disaster marginalization in the aftermath of disasters. It connects with broader emancipatory literature on dilemmas involved in empowerment "from the outside." In a future of potentially harsher climate-related disasters and increased social vulnerability for certain communities, this book contributes to a full and nuanced understanding of community empowerment and vulnerability reduction.

This book will be of interest to sociologists, anthropologists, geographers, political scientists, and urban studies researchers, as well as undergraduate and postgraduate students interested in disaster management, disaster risk reduction, social vulnerability, community empowerment, development studies, local studies, social work, community-based work, and emancipatory theory.

**Sara Bondesson**, Associate Professor at the Swedish Defence University, is a political scientist interested in disasters, identity and power. Sara combines theorizing with ethnographic methods, since she believes normative, emancipatory, and transformative theories are best explored from the ground up. Apart from the Swedish Defence University, she is also affiliated with the Centre of Natural Hazards and Disaster Science. In teaching future political scientists and crisis managers she works with participatory and scenario-based methodologies and usually mix it up with storytelling or improvisational methods.

## Routledge Studies in Hazards, Disaster Risk and Climate Change

**Series Editor:** Ilan Kelman, Professor of Disasters and Health at the Institute for Risk and Disaster Reduction (IRDR) and the Institute for Global Health (IGH), University College London (UCL)

This series provides a forum for original and vibrant research. It offers contributions from each of these communities as well as innovative titles that examine the links between hazards, disasters and climate change, to bring these schools of thought closer together. This series promotes interdisciplinary scholarly work that is empirically and theoretically informed, with titles reflecting the wealth of research being undertaken in these diverse and exciting fields.

**Why Vulnerability Still Matters**
The Politics of Disaster Risk Creation
*Edited by Greg Bankoff and Dorothea Hilhorst*

**Health, Wellbeing and Community Recovery in Fukushima**
*Edited by Sudeepa Abeysinghe, Claire Leppold, Alison Lloyd Williams and Akihiko Ozaki*

**Slow Disaster**
Political Ecology of Hazards and Everyday Life in the Brahmaputra Valley, Assam
*Mitul Baruah*

**Empowerment and Social Justice in the Wake of Disasters**
Occupy Sandy in Rockaway after Hurricane Sandy, USA
*Sara Bondesson*

For more information about this series, please visit: www.routledge. com/Routledge-Studies-in-Hazards-Disaster-Risk-and-Climate-Change/book-series/HDC

# Empowerment and Social Justice in the Wake of Disasters

## Occupy Sandy in Rockaway after Hurricane Sandy, USA

**Sara Bondesson**

Routledge
Taylor & Francis Group

LONDON AND NEW YORK

First published 2023
by Routledge
4 Park Square, Milton Park, Abingdon, Oxon OX14 4RN

and by Routledge
605 Third Avenue, New York, NY 10158

*Routledge is an imprint of the Taylor & Francis Group, an informa business*

*British Library Cataloguing-in-Publication Data*
A catalogue record for this book is available from the British Library

*Library of Congress Cataloging-in-Publication Data*
A catalog record has been requested for this book

ISBN: 978-0-367-43719-0 (hbk)
ISBN: 978-1-032-35894-9 (pbk)
ISBN: 978-1-003-00527-8 (ebk)

DOI: 10.4324/9781003005278

Typeset in Times New Roman
by Newgen Publishing UK

This book is dedicated to my son, Stig.

This book is dedicated to my son, Stig.

# Contents

# Acknowledgments

Writing this book has been a winding journey and many people have been there along the way. The team at Routledge deserve a great thank you; Ilan Kelman, Faye Leerink, Nonita Saha, Esha Gupta, and Thara B Kanaga, as well as the two anonymous reviewers. I am also grateful to my colleagues in the academic departments I have belonged to during the years. My fellow crisis and leadership researchers and the fine folks of the Gender, Peace and Security group, at the Swedish Defence University, have been invaluable sources of support and inspiration. I also want to give thanks to Uppsala Department of Government, especially the Qualitative Methods seminar and the Gender and Politics seminar, and the interdisciplinary spirit of the Centre of Natural Hazards and Disaster Science. My outmost gratitude goes to Elisa Viteri who has been of invaluable assistance and guidance in the late drafting of the book. I could not have asked for a more competent and insightful reader: thank you Elisa, you made this book better. All mistakes and mishaps are mine though. I also want to extend my deepest appreciation to all interview persons and field contacts. The fact that some of you not only took the time to sit down with me to share your stories and experiences, but helped me navigate the context of Rockaway and New York City, fed me during Thanksgiving dinners, and invited me into your homes is remarkable. This book had not been possible without you.

# Abbreviations

CBA      Community Benefits Agreement
FDNY     Fire Department New York
FEMA     Federal Emergency Management Agency
NYC       New York City
NYCHA    New York City Housing Authority
NYPD      New York Police Department
OWS      Occupy Wall Street
SIRR       NYC Special Initiative for Rebuilding and Resiliency
SRO       Single Residence Occupancy
SRO       Hotel Single Room Occupancy Hotel

# 1 Shaking Things Loose

## Meeting Occupy Sandy in Rockaway

In May 2013, I found myself in a run-down church in Rockaway, Queens, New York City (NYC), six months after Hurricane Sandy had brought havoc to this oceanfront peninsula. The church was scene for a meeting between mostly white, mostly young Occupy Sandy activists and storm-affected residents, mostly people of color, from some of Rockaway's politically, socially, and economically marginalized communities.

Outside the church, the scars of the hurricane were still visible: the surrounding houses with their boarded-up windows, no repairs in sight, the sleepy subway track bridges, still of no use. The shattered boardwalk left in pieces along the beach. Residents still talked about the storm as "the tsunami," remembering the giant wave that came crashing in on October 29, 2012. They had seen their houses destroyed, their electricity cut, their connection to the world lost. They had waded through water mixed with sewage sludge, venturing out to find food as their groceries rotted in their fridges. They had lost their jobs. Six months later, some still lived in temporary housing, others were still unemployed and yet others suffered from mold-induced health problems and substance abuse issues triggered by stress. But they were at least out of the deepest woods and were starting to think about the ways to rebuild and hopefully come again, stronger than before.

As Hurricane Sandy made landfall on the shores of NYC, Occupy Sandy had turned their dormant Occupy Wall Street networks into a web of relief work, mobilizing roughly 60,000 activist volunteers to support local neighborhoods across the City in a struggle to overcome the social, economic, and political disaster of Hurricane Sandy. Occupy Sandy activists channeled large quantities of goods and supplies to NYC's hardest-hit areas. Crews of Occupy activists had taken on the heavy, difficult, and dirty job of making sure that disaster-affected

DOI: 10.4324/9781003005278-1

communities received shelter, food, and warmth. Making do with what they had, they assisted storm-affected people and provided crucial human connection in a disaster that had cut off power and communication to whole neighborhoods. Mutual support was an integral part of the Occupy ideology, and the activists were well aware of the dysfunctional power dynamics that can come into play between saviors and helped.

As a feminist activist and newly minted PhD student in political science with a focus on crisis management, the case of Occupy Sandy was new and very exciting. I was curious to see it all manifested. Making my way out to Rockaway that afternoon in May, I could not wait to experience the antithesis of bureaucratic government agencies operating out of command-and-control logics. What I observed, however, was far from my expectation.

The light breeze from the ocean mixed with the late afternoon sun when I stepped into the church and made my way up the shaky stairs to the second floor. With roughly 50 people of all colors present, the room was buzzing with life. Rockaway residents, Occupy Sandy activists, and a bunch of children were eating food, talking with each other, shouting across the tables, mingling, coming and going as they pleased. I was handed some food and sat down to talk to a white man, a few years older than me, who immediately asked me what "my politics" were. I mumbled something about having a research interest in just recovery processes. This seemed to please him and he embarked on a long story about how he had come to be politically conscious. When the meeting started after an hour or so, a white activist in her mid-20s, a scarf haphazardly thrown round her neck, explained that the Occupy Sandy hub wanted to mobilize the Rockaway community and empower people to take the lead in the long-term recovery work that laid ahead. She then initiated a discussion exercise meant to tease out "allies in the wider community." Thrilled and attentive, I took a seat in the somewhat messy circle of chairs that had formed around her.

To my surprise, two black male residents in their 50s immediately interrupted her and started to ask questions about what kind of organization Occupy Sandy was, and where the money was coming from. The woman answered something short and vague and then tried to go back to the exercise, but the residents would not let her. One of the men, a burly man dressed in an over-sized gray hoodie, stood up: "When you talk about 'we', who are you talking about? Because 'we' [*pointing to the other residents surrounding him*] are not even in the same time-zone as 'you'!" [*pointing to the activists*].

I held my breath and tried to make myself a fly on the wall. I felt increasingly uncomfortable, being a white young-ish woman myself, a scarf wrapped around my neck. Easily mistaken for an Occupy Sandy activist.

The meeting dragged on and issues of finances and transparency kept coming up in increasingly frustrated iterations. Residents insisted on asking questions such as: "Who decides here?" "Who is the leader?" and "Who signs off on the checks?" Some of the activists tried to steer the discussion away from these questions by saying: "There is no organization here, this is a non-hierarchical space, so there are no strict answers to that." When the residents kept repeating their questions, the activists said: "Now you have to step down and let the meeting happen" or "Your question is not genuine so it does not need a genuine answer." A male activist finally lost his temper and, while some of the female activists discretely rolled their eyes, aggressively shouted: "We are not about authority!"

When the meeting ended, late at night, and everyone spilled out on the broken sidewalk, I spoke to one of the most confrontational residents, Jemar, a young and well-dressed black man. Jemar turned out to be the leader of a local organization in Rockaway. When I asked him why he had challenged the activists, he said: "I have seen so many organizations come and go, and they gain the trust of the people here and then they leave. We felt there were things that needed to be sorted out."

There was a certain irony to the situation, I thought. Here was a bunch of ideologically motivated young people devoting their energy to building a nonhierarchical emancipatory project in the aftermath of a hurricane that had struck unevenly across their city. They wanted to set in motion a political awakening process that would empower residents to take the lead in the long-term recovery from this storm. Yet, they stumbled upon resistance from the very same residents that they wanted to empower, a resistance so spirited that it completely shut down the agenda of the meeting.

The situation speaks to a larger puzzle that runs like a thread through much of the literature on emancipation in social justice spaces. Empowerment is understood as a process where people gain control over their lives, often through participation in arenas of social change. As discussed by, for example, Andrea Cornwall (2002, 2016), Gada Mahrouse (2014), and the late Iris Marion Young (2000), social justice spaces are often initiated by people who are not themselves part of the communities they set out to empower. Yet, true empowerment is believed to come from within the community itself. It was fascinating,

I thought, to see some of the tensions that this puzzle may bring about, played out in a disaster context.

I was also thinking about the timing. We were six months after the storm, and the immediateness of the relief period was winding down. I could sense that this organizational hub was about to shift into something more long term, the recovery period. Yet, it was obviously going through some growing pains that I suspected had something to do with the potentially messy transition from one phase to another. Perhaps, as Cornwall (2016) so elegantly puts it, the empowerment process that was taking place traveled along a meandering path, through terrain "pitted with thorny thickets, fast-flowing rivers, mud and marshes, and along paths that can double-back on themselves, meander on winding side-routes and lead to dead-ends, as well as opening up new vistas, expanding horizons and extending possibilities" (p. 345).

Over the next three years, I spent as much time as possible in Rockaway, tracing Occupy Sandy and their work with Rockaway residents. Although a political scientist by training, I borrowed disaster anthropology's key stone methodology – ethnography – in the hope of unearthing activists and residents' fluctuations between collaboration and conflict, as well as the winding routes of possible empowerment in the shift from immediate relief to long-term recovery. I lived in Rockaway, where I shadowed and interviewed residents, activists, and other community organizers. I also studied their work up close, through participatory observations of rallies, demonstrations, meetings, and parties. Learning from residents and activists, listening to their stories, and observing their collaborations, practices, and activities, I worked through a set of questions that left me no peace: Why were the residents so angry? What is it like to be someone whom others feel the urge to save, to empower? What is it like when those who come to empower you are younger, whiter, perhaps richer and less damaged by the disasters you have seen and felt, not only as sudden-onset storms but also in your daily life? And also, what kind of strains can we expect to see in the move from immediate relief to long-term recovery and what implications does the transition have for the meandering paths of empowerment? This book is an attempt at an answer.

## Shaking Things Loose – Disasters as Windows of Social and Political Change

On March 31, 2020, Rebecca Solnit was interviewed in the podcast *On Point*. In her calm, almost soothing voice, Solnit stated: "Every disaster shakes loose the old order: The sudden catastrophe changes the rules and demands new and different responses, but what those will be are the subject of a battle." Solnit's words ties into the question "what happens after?" After the disaster strikes, after the conflict flares up, as societies transition back, or move into something new. Can a disaster be an opportunity to move away from the problems of inequality that characterize many societies of today, or are we doomed to fall back into the same patterns? And most importantly, which battles will mark the transition? These very questions are the starting point of this book, which focuses on how Hurricane Sandy, as it tore Rockaway into pieces, also triggered a grassroots political movement that aimed to empower low-income communities of color to shake loose their lived experiences of disempowerment and marginalization and take matters into their own hands.

This post-disaster timeline can roughly divide into relief and recovery (Fothergill & Peek, 2004). Relief often includes the first days and weeks, up to a few months. Here focus is on saving lives and homes, organizing evacuations, securing infrastructure, and making sure that people's basic needs are met. Recovery covers approximately the one-year period following the disaster, in which schools, roads, homes, and businesses are to reopen, and people are to find a way back to sustain their livelihoods. Depending on the level of vulnerability, however, or the scale of the hazard, recovery can extend into several years. Sometimes it just never happens, but people are thrown into perpetual states of despair.

Disaster relief, although always contextually and situationally varied, show certain similarities and patterns, as showed by anthropologists Anthony Oliver-Smith and Susanna Hoffman in their landmark volume *The Angry Earth* (1999). No matter what sort of hazard, no matter the level of social vulnerability that structured the effects of the hazard, large-scale disasters tend to bring about a need for shelter, warmth, safety, water, food, and community. Furthermore, disaster survivors often unite and merge in new ways, at least in the acute stages after a disaster. New forms of relationships can emerge, and new affiliations be formed. Oliver-Smith and Hoffman (1999) speak of the "spontaneous social solidarity that temporarily enables people to put aside self-interest and come together in common effort" (p. 156). Sancha Medwinter (2021), in studying nongovernmental organization (NGO) volunteer–resident relations in Rockaway after the storm, suggests an emotive dimension that may come into play for trust building and social bonding between people in acute disaster situations. She states that the context of acute crisis may catalyze bonding, since the gravity of loss and survival sets in motion affective attachment between responders and survivors. Through such reinvention of social relations, power dynamics may be altered and potentially even reversed.

Yet fragmentation often occurs as the relief stage moves into long-term recovery, and as people start searching for meaning and explanations for what happened. Disasters, as Oliver-Smith and Hoffman (1999) point out, stir conflict because they bring about the potential for change by shining light on pre-disaster conditions in need of alteration. There are studies showing how communities become politicized as disasters make inequalities visible and fortify ongoing political struggles (Green, 2008).

Some disaster situations have been shown to set forth open policy dialogues, create an enhanced sense of ethnic identity, or even play a role in post-disaster elections (Fothergill et al., 1999). After Hurricane Andrew struck the United States in 1992, new community projects improved poor neighborhoods, and after the Loma Prieta earthquake in 1989, affordable housing was created for low-income families (Fothergill & Peek, 2004). Another example is the Chilean earthquake of 1939, which killed 30,000 people and triggered accelerated policy change (Pelling & Dill, 2010). Thus, disasters can function as critical junctures and contribute to political change toward the emergence of more egalitarian policies in the post-disaster phase (Pelling & Dill, 2010). State failures to respond accurately to disasters may create temporary power vacuums that open up for contending civil society actors working for systemic change (Boin et al., 2008; della Porta & Dianni, 2006; Hannigan, 2012; Pelling & Dill, 2010). Whether such change occurs, however, is an empirical question (Oliver-Smith & Hoffman, 1999).

Other studies paint a gloomier picture. Macroeconomic analysis shows that disasters can lead to significant capital influx, but that the inflows benefit affluent social groups. Interventions that seek to challenge poverty and powerlessness are likely to be resisted by privileged groups. Moreover, external assistance in disaster situations tends to reinforce rather than undercut existing social structures (Dynes, 2002; McEntire, 1997).

## Barking Up the Wrong Tree – Crisis Management's Inability to Fix Social Vulnerability

A steady stream of disasters has plagued the world over the past 20 years. The 2004 Indian Ocean Tsunami that devastated the coastal regions of Southeast Asia; hurricanes that strike the Caribbean and the United States on a yearly basis; the overwhelming effects of the Haiti earthquakes in 2010 and 2021; the 2011 Pakistani floods; Typhoon Haiyan's destruction of the Philippines in 2013; Hurricane Matthew that took thousands of lives in 2016 and in 2021; or Hurricane Ida that struck the United States with devastating results. But as Ilan Kelman (2020) writes, "the tornado, the earthquake, the tsunami are not to blame" (p. vii). The real disasters are the deaths and injuries, the loss of homes, of livelihoods, and the failure to support the affected people from the devastating natural forces (Kelman, 2020). Disasters are processes that lead to a potentially destructive event in which a natural or technological agent affects a population that lives in socially produced conditions of vulnerability (Oliver-Smith & Hoffman, 1999). Focus then, is less on the results of natural hazards as such, and more on how ongoing social orders structure vulnerability to these hazards.

Disasters hit some people harder than others, and the root causes for unequal distribution of risk and vulnerability are found in the very fabric of society. Research on social vulnerability shows how factors deeply rooted in our societal structures lead to varying levels of vulnerability for different social groups (Abramson et al., 2015; Bankoff et al., 2004; Enarson & Morrow, 1998; Fothergill et al., 1999; Fothergill & Peek, 2004; Jones & Murphy, 2009; Thomas et al., 2013; Tierney, 2014; Wisner et al., 2004). Structural inequalities are at work when race, gender, class, ability, or any other social marker conditions a person's legal status, their educational possibilities, and their access to political power (Young, 2000). As Lori Peek (2019), director of the Natural Hazards Center points out, people die in disasters because of:

economic inequality and generational poverty that traps families in shoddily built homes and children in unsafe schools. They die because of corporate greed and voter apathy. They die because of political calculations and climate change denialism. They die because of poor land use planning and a lack of building code adoption and enforcement. They die because they've been pushed into low-lying areas where rapidly disappearing land meets the dangerous sea.

The late sociologist Zygmunt Bauman (2011) claimed that the likelihood for socioeconomically marginalized people to become victims of disasters is the most "disastrous among the many problems humanity may be forced to confront, deal with and resolve in the current century" (p. 9). Moreover, the very same problems of inequality that put some people in harm's way also manifest in communities' lack of power and political voice.

Yet, even if disaster scholars, especially those trained in anthropology or sociology, agree that disasters are explicit manifestations of ongoing structural inequalities, most crisis management practitioners (as well as some crisis management scholars) operate according to logics that make it hard to identify and rectify inequalities. The paramilitary bureaucratic command-and-control approach to crisis management that grew out of the Cold War era of civil defense promotes strict, rigid, and bureaucratic norms with centralized and hierarchical decision-making (Gardner, 2013; Neal & Phillips, 1995; Schneider, 1992; Whittaker et al., 2015). Predefined objectives, formal structures, clear division of labor, and policies and procedures that steer all activities are part and parcel of many crisis management organizations (Schneider, 1992). Such rationalistic frameworks are based on a utilitarian logic and the assumption that crises follow linear processes of detection, preparation, containment, recovery, and learning (Branicki, 2020). Yet, as I discuss elsewhere, these short-term, techno-managerial logics make the analysis of structural inequalities unintelligible and silences any possible political solutions that could alter them (Bondesson, 2019). In essence, to expect mainstream crisis management practice to tackle social vulnerability is barking up the wrong tree.

## What Does Occupy Sandy in Rockaway Bring to the Table?

The story of Occupy Sandy is another crisis management story altogether – a story of a relief and recovery process mired in social justice, involving the most affected and least heard, aiming to empower them to take the lead and alter the unjust life circumstances that had put them in harm's way in the first place. Occupy Sandy's lineages to the ideological underpinnings of the wider Occupy movement, expressed through the call for mutual aid, made their relief activities completely different from the kinds of operating logics often seen in traditional crisis management. Occupiers, drawing on anarchist discourse and rejecting hierarchical organization, put forward horizontal networks based on what Juliane Reinecke calls prefigurative organizing (2018). Prefigurative organizing is the enactment of a desired future meant to disrupt the reproduction of entrenched inequalities. It is perhaps best captured in the simple statement: "be the change you want to see in the world" (Reinecke, 2018, p. 1299). Prefigurative organizing addresses inequalities by directly shifting power relations on a micro-political level – for example, through horizontal decision-making – with the end goal of transforming inequalities. In this way, ends and means are aligned, as organizing becomes both an instrument and a way of expressing the desired state. Participants' everyday practices involve imagining and experimenting with alternative ways of doing and relating to others, in "the here and now" (Reinecke, 2018, p. 1302). The case of Occupy Sandy thus represents an actor deeply aware of (and conscious not to reproduce) existing inequalities between helper and helped. Yet along the way, something had gone sour, as was evident in my first vivid encounter with Occupy Sandy in Rockaway.

The case of Occupy Sandy in Rockaway sits smack in the intersection between crisis management and social justice. Hence, this book speaks both to emancipatory theorists interested in social movements' organizing, as well as researchers focused on crisis and disaster management. Disaster sociologists, disaster anthropologists, but also urban studies scholars and cultural geographers may find this book useful. Finally, this book particularly contributes to the crisis management research field of emergent groups.

Generally, emergent groups are spontaneous groups of citizens and relief workers that arise in response to emergencies where established norms of behavior, roles, and practices come into flux because of the severity and uncertainty of the situation. Emergent groups have been theorized and researched from various angles, especially in regard to their effectiveness and their possibilities to channel citizens' altruism, as well as the challenges they pose for established emergency management organizations in relief situations (Drabek & McEntire, 2003; Dynes, 1970, 1987; Emmanuel, 2006; Gardner, 2013; Stallings & Quarantelli, 1985, 2015; Voorhees, 2008; Whittaker et al., 2015). However, with a few exceptions (see Carlton, 2015; Seana & Fothergill, 2009; Solnit, 2009; Twigg & Mosel, 2017), little attention has been directed to the role that emergent groups may play in empowering marginalized and vulnerable communities in the long run, as relief turns into long-term recovery.

On the other hand, empowerment is at the very core of the emancipatory literature, an umbrella term for studies on citizen participation, social movements, grassroots organizing, and democratic theory (Ahmed, 2004; Bacchi, 1996; Benhabib, 1996; Butler et al., 2016; Chavis, 2001; Choudry et al., 2012; Cornwall, 2002; Cornwall & Schatten Coelho, 2007; Crenshaw, 1991; Dovi, 2009; Holvino, 2008; hooks, 2010; Houten & Jacobs, 2005; Jordan-Zachary, 2007; Jung, 2003; Kruks, 2001; Phillips, 1995; Pilisuk et al., 1996; Scott & Liew, 2012). As rich and sometimes even mind-blowing this literature is (who has ever read the late bell hooks, Sarah Ahmed or Judith Butler without feeling that the world comes before you anew?), not many empirical studies focus on social justice spaces in post-disaster processes. Social justice movements do not usually engage in direct aid. How can we make sense of an empowerment process that was triggered by a disaster, and that involved people marginalized before and by it? How does a disaster change the dynamics of social justice organizing? The work of Occupy Sandy in Rockaway was a micro-cosmos well suited for exploring these questions.

Occupy Sandy activists involved Rockaway residents, primarily people of color, with the aim of empowering them by letting them lead the recovery process according to their own priorities as they moved forward from the immediate effects of the storm. Participants spanned social identities and hierarchies in a thought-provoking mix of social, educational, economic, and racial backgrounds. The Occupy Sandy activists were mostly white young people, mostly educated, mostly not affected by the storm, and in possession of organizational skills as well as economic funds that they controlled. In turn, Rockaway residents were mostly low-income people of color struggling to get back on their feet after a storm that had ripped their neighborhood to pieces. They had little experience of the type of social justice work in which they were invited to take part. Tracing the shifts from relief to recovery, this book furthers a better understanding of the internal dynamics in the transition from direct aid to long-term efforts, especially on the looming effects of disasters on issues of power and influence. In sum, by studying Occupy Sandy's work in Rockaway up close, we can learn new things about post-disaster empowerment in micro-political shifts from direct relief aid to long-term organizing.

## Mixing It Up – Living and Breathing the Field

If the story of Occupy Sandy in Rockaway sits in the intersection between crisis management scholarship and emancipatory scholarship, something of the sorts goes for me as well. A feminist activist turned political scientist researcher, I work in a crisis and disaster management research environment. I am constantly negotiating the ambiguities that come along with being in such an overlap. Crisis and disaster management researchers are often part of the network of experts that shape the policy and practice of crisis and disaster management. Sometimes their perspectives exhibit the same depoliticized, utilitarian, rationalistic, and techno-managerial logics that undergird crisis management practice. This noise can be hard to cut through, especially for someone who devotes her time to rather complex issues of power, identity, and inequality in relation to disasters and crises – issues difficult to summarize in neat bullet points for policy action. At times I have tried, as Swati Parashar (2021) so elegantly puts it, to "undiscipline myself" from the crisis management discourse. In so doing, the universe of feminist, critical, and post-colonial thinking has provided me with a home, albeit a confusing and challenging one at times.

During this time, I have also turned to disaster sociology and disaster anthropology, fields that view disasters not as time-delineated events to be managed, but as violent expressions of underlying historical structures of inequalities. In studying Occupy Sandy's shift from relief to recovery and the fluctuations of collaboration and conflict between activists and Rockaway residents, I made use of disaster anthropology's central methodology, ethnography, which lends itself well to studies of complex micro-practices. This method also enabled me to explore how people understand and navigate their contexts within micro-political, ever-changing, organizationally fluid social justice movements (Balsiger & Lambelet, 2014; Gustafsson & Johannesson, 2016; Oliver-Smith & Hoffman, 1999; Wolford, 2006). The book is based on 11 months of field-work spread out over three years from 2013 to 2016, 44 semi-structured interviews, 8 participatory observations, 12 field study visits to Sandy-related events in and out of Rockaway or study visits at organizations presenting their relief and recovery work, one month of volunteer work in a local community organization in Rockaway, and numerous informal conversations and encounters with residents, activists, local politicians, community organizers, crisis management practitioners, and other scholars doing research on Rockaway. All of this helped me capture both the inner dynamics of Occupy Sandy, as well as the political, social, and economic context of Rockaway, including the effects of the storm. All of these are factors that played into what took place within Occupy Sandy's relief and recovery empowerment efforts.

Fieldwork shed light on the at times complicated relations between residents and Occupy Sandy activists. It also gained me more of an insider status, which helped in getting further access and interviews. As a white Swedish woman in my 30s, my positionality in relation to the people I met shifted, sometimes several times a day. I blended in rather well among the white Occupy Sandy activists but stood out when I went for meetings in the eastern parts of Rockaway, where I would often be the only white person in sight. I navigated some social codes with help of clothing, manner, and language to be less intrusive with my presence. Sometimes I just had to accept that I was seen as alien. On a regular day I could have a scheduled interview at the NYC Planning Department to talk about urban renewal in Rockaway after the storm, only to leave the building, take off my blazer jacket, ruffle my hair a little, and walk a few blocks up to Wall Street to participate in "Flood Wall Street," an aggressive anti-capitalist march where activists were arrested *en masse* (and where I would make sure to stay well on the right side of the fence so as not to get arrested myself).

Most of my time however, I spent in Rockaway. I rented a room in the home of Caroline and Mary, a progressive couple in their 60s, who lived in a small house they lovingly called "their bungalow." Located in an affluent, mostly white, Jewish and Irish community on Rockaway's western side, the house was only a stone's throw away from the sandy white beach. Caroline and Mary's house was small but homely, with old wooden furniture, a large inviting couch in the living room and whimsical collections of shells decorating the windowsills. In ethnographic field studies, anyone and anything you encounter or observe – even if you are simply out buying groceries for dinner – is a potential source of data. This means that you are constantly trawling; constantly navigating every social setting you are in, which in turn means that your head is pretty much on fire. All the time. Caroline and Mary's home became a refuge, a safe space where I could tuck myself away from the never-ending data-collection mode of ethnographic field studies. When I was not attending rallies or meetings, or out interviewing people, or head down in the latest issue of *The Wave* – Rockaway's local newspaper – I spent time with Caroline, listening to her stories of Rockaway and the storm. I relied on her to help me understand the area, the politics, and the people of Rockaway. She was an important cultural interpreter and soon became a close friend.

In approaching people for interviews or informal conversations, I would introduce myself, my university affiliation, and briefly describe my research project. I then asked if I could talk to them about their experiences while also informing them that I would use pseudonyms to keep their names confidential and that they could withdraw their participation at any time. Out of the 44 interviews, 32 were respondent interviews and 12 were informant interviews. Informant interviews were made with organizers from community-based organizations, social workers involved in servicing residents affected by the storm, politicians representing the district at the local, city, and state level, urban development experts, private developers, city government officials, and state department officials. Through these interviews, I familiarized myself with Rockaway, the storm, and its effects on vulnerable groups, as well as the plethora of different community organizations active in the area. I also learnt more about urban renewal in the aftermath of the storm, an issue in which Occupy Sandy was involved.

The 32 respondent interviews were selected through a theoretical sampling intended to shed light on the inner dynamics of Occupy Sandy. I wanted to capture the different perspectives of those who were actively involved, both among residents and activists. To get a richer and more nuanced understanding, I also interviewed residents who had dropped out, as well as those who were ideologically aligned with Occupy Sandy but had chosen not to engage. The respondent interviews were semi-structured, exploring the interviewee's social, economic, and educational backgrounds, self-ascribed identity, general political outlook, and whether and how they were affected by the storm. We also talked about their understanding of the internal dynamics, meeting techniques, potential conflict, and identity-based power relations within Occupy Sandy. The questionnaire developed over time to include questions about specific situations of tension, as these were revealed in participatory observations.

I also conducted participatory observations of public meetings, workshops, film screenings, demonstrations and parties, focusing on the environment of collaboration and conflict. In these observations, I took part in discussions and exercises, although I tried to keep my own remarks to a minimum. I also noted down what was being said during and immediately after the meetings and took note of rhetoric and communicative styles. I observed the ins and outs of meeting techniques, the facilitation of exercises, how emerging leaders were supported and encouraged while also being on the lookout for covert or overt conflicts and tensions. Apart from resulting in data, the participatory observations made interviews run more smoothly because they helped me get to know the residents and activists. This made interviewees more comfortable in sharing their perceptions, frustrations, and experiences. Observations also functioned as impetus into the interview questions. I would ask about particular moments of tensions that I observed to get more information on how these moments were experienced by the interviewee. In observing strained interaction, it was easier for me to identify silent frustrations among interviewees and subtly probe them on these issues.

Apart from observations and interviews, I volunteered in a small local community organization that opened in October 2012. Its director was a local resident with a mission to bring employment opportunities to the socioeconomically marginalized communities of Rockaway, with an emphasis on green sustainable technologies. The office was located in an unassuming one-story shopfront building. Next to a small deli, it was surrounded by a number of single-residence occupancies and by apartment buildings that housed low-income renters. The door to the organization was often open, letting in the brisk winds from the Atlantic Ocean together with a steady line of people that came and went. Spending my time there, I came to meet Rockaway residents of all flavors. It was mostly men of color in need of assistance with job hunting, vocational training or food stamps, but also visitors to the area curious about Rockaway's dealings with the hurricane and occasionally, other community organizers who came in to chat or coordinate with the director.

Besides real-life encounters, I followed a few social media groups and accounts. In these, I found relevant input into how different Rockaway communities and activist circles reasoned with issues of importance. Since I followed processes that were playing out while studying them, it was important to track social media activity when I was away from the field. To follow Facebook pages or Twitter accounts was a good way of keeping an ear to the ground.

I also kept a daily journal, which served different purposes. I jotted down key words or phrases while I was in the field and took note of spontaneous observations or conversations. I also used the journal as a way of keeping track of the more purposeful participatory observations I conducted and wrote down descriptions of physical locations or interactions between people and their behavior and nonverbal communication. I also used it for writing down emerging ideas for alterations of research questions and documented the self-reflective tenets of the fieldwork – the reactions from interviewees, my own frustrations, hopes and fears, and other emotional responses that came up during the course of fieldwork.

## Outline of the Book

Following this introduction in Chapter 1, Chapters 2 and 3 discuss the theoretical and empirical problems that motivate and set the directions of this book: inequality, vulnerability, and empowerment in light of disasters. Chapter 2 gives an overview of social vulnerability to disasters and illustrates how it is connected to structural marginalization based on race, class, gender, or other social power orders. I also discuss how marginalization of low-income communities of color ties into political marginalization so that the most affected by disasters are also the least heard.

In Chapter 3, I wed two distinct literatures to each other – the emancipatory literature and the literature on emergent groups – in order to gauge what empowerment is in a disaster context. I also discuss a dilemma of empowerment, namely that empowerment processes are often run by people who are not themselves part of the marginalized communities they wish to empower; yet genuine empowerment is believed to come from within the community itself. I end by elaborating on a few tensions that this dilemma brings about in a post-disaster setting.

In Chapters 4 through 6, I present the empirical story of the book. I set the scene in Chapter 4 by describing the peninsula of Rockaway, its political, social, and economic marginalization and how these conditions of underlying vulnerability were exacerbated by Hurricane Sandy. In Chapter 5, I detail the work of Occupy Sandy in Rockaway during the relief process, where Occupy Sandy's attention to inclusion, autonomy, and horizontality amplified and strengthened local relief activities. I focus on the micro-political power dynamics that came into play between activists and residents and whether and how residents of Rockaway were empowered by their participation in the Occupy Sandy network.

The empirical narrative then moves toward the recovery period in Chapter 6. During the relief period, Occupy Sandy and their work with Rockaway residents was a story of heartwarming collaboration across social and political divides. However, this chapter shows how the journey from relief to recovery brought brewing tensions that grew into outright conflicts. While some residents felt empowered by their collaboration with the movement, others harshly criticized Occupy Sandy activists for reproducing the same power unbalances that they were trying to undo.

In the final Chapter 7, I summarize the story of Occupy Sandy in Rockaway and propose a concept to capture the conundrum of this process: "the savior trap." The savior trap connotes how activists, although deeply aware of their own privileges, may end up replicating some of the same power imbalances they set out to alter. In essence: we cannot save, but we cannot *not* save. I then situate the ongoings of this particular social justice space in larger oppositional struggles, at a time when we are witnessing harsher disasters as climate changes and social, political, and economic inequalities widen. In concluding, I discuss another kind of trap I found myself in when writing this book and consider the implications of my study.

# References

Abramson, D., Van Alst, D., Merdianoof, A., Piltch-Loeb, R., Beedasy, J., Findley, P., … Tobin-Gurley, J. (2015, April 1). *The Hurricane Sandy Place Report: Evacuation Decisions, Housing Issues and Sense of Community* (Briefing Report 1). The Sandy Child and Family Health Study. Rutgers University School of Social Work, New York University College of Global Public Health, Columbia University National Center for Disaster Preparedness, & Colorado State University Center for Disaster and Risk Analysis.

Ahmed, S. (2004). Declarations of whiteness: The non-performativity of anti-racism. *Borderlands*, 3(2). Retrieved from https://research.gold.ac.uk/id/epr int/13911/

Bacchi, C. L. (1996). *The Politics of Affirmative Action: 'Women', Equality and Category Politics*. London: Sage Publications.

Balsiger, P., & Lambelet, A. (2014). Participant observation. In D. della Porta (Ed.), *Methodological Practices in Social Movement Research* (pp. 144–173). Oxford: Oxford University Press.

Bankoff, G., Frerks, G., & Hilhorst, D. (2004). *Mapping Vulnerability: Disasters, Development and People*. London, UK: Earthscan.

Bauman, Z. (2011). *Collateral Damage: Social Inequalities in a Global Age*. Cambridge, UK: Polity Press.

Benhabib, S. (1996). *Democracy and Difference: Contesting the Boundaries of the Political*. Princeton: Princeton University Press.

Boin, A., McConnell, A., & 't Hart, P. (2008). *Governing After Crisis: The Politics of Investigation, Accountability and Learning*. Cambridge, UK: Cambridge University Press.

Bondesson, S. (2019). Why gender does not stick: Exploring conceptual logics in global disaster risk reduction policy. In C. Kinnvall & H. Rydström (Eds.), *Climate Hazards, Disasters, and Gender Ramifications* (pp. 88–124). London: Routledge.

Branicki, L. J. (2020). COVID-19, ethics of care and feminist crisis management. *Gender, Work and Organization*, 27(5), 872–883.

Butler, J., Gambetti, Z., & Sabsay, L. (Eds.). (2016). *Vulnerability in Resistance*. Durham and London, UK: Duke University Press.

Carlton, S. (2015). Connecting, belonging: Volunteering, wellbeing and leadership among refugee youth. *International Journal of Disaster Risk Reduction*, 13, 342–349.

Chavis, D. M. (2001). The paradoxes and promise of community coalitions. *American Journal of Community Psychology*, 29(2), 309–320.

Choudry, A., Hanley, J., & Shragge, E. (2012). *Organize: Building from the Local for Global Justice*. Oakland, CA: PM Press.

Cornwall, A. (2002). *Making Spaces, Changing Places: Situating Participation in Development* (IDS Working Paper 170). Brighton, UK: Institute of Development Studies.

Cornwall, A. (2016, April). Women's empowerment: What works? *International Development*, 28(3), 342–359.

Cornwall, A., & Schatten Coelho, V. (2007). *Spaces for Change? The Politics of Citizen Participation in New Democratic Arenas*. New York, NY: Palgrave Macmillan.

Crenshaw, K. (1991). Mapping the margins: Intersectionality, identity politics, and violence against women of color. *Stanford Law Review*, 43(6), 1241–1299.

della Porta, D., & Diani, M. (2006). *Social Movements: An Introduction* (2nd ed.). Hoboken, NJ: Wiley.

Dovi, S. (2009). In praise of exclusion. *The Journal of Politics*, 71(3), 1172–1186.

Drabek, T. A., & McEntire, D. A. (2003). Emergent phenomena and the sociology of disaster: Lessons, trends and opportunities from the research literature. *Disaster Prevention and Management: An International Journal*, 12 (2), 97–112.

Dynes, R. R. (1970). *Organized Behavior in Disaster*. Lexington, MA: Health Lexington Books.

Dynes, R. R. (1987). The concept of role in disasters. In R. R. Dynes, B. de Marchi, & C. Pelanda (Eds.), *Sociology of Disasters: Contribution of Sociology to Disaster Research* (pp. 71–103). Milano, Italy: Franco Angeli.

Dynes, R. R. (2002). *Disaster and Development, Again*. Newark, DE: University of Delaware.

Emmanuel, D. (2006). Emergent behavior and groups in postdisaster New Orleans: Notes on practices of organized resistance. In Natural Hazards Center (Ed.), *Learning from Catastrophe: Quick Response Research in the Wake of Hurricane Katrina* (pp. 235–261). Boulder, CO: Natural Hazards Center, University of Colorado.

Enarson, E., & Morrow, B. H. (1998). *The Gendered Terrain of Disaster: Through Women's Eyes*. New York, NY: Praeger Publishers.

Fothergill, A., Maestas, E. G. M., & DeRouen Darlington, J. (1999). Race, ethnicity and disasters in the United States: A review of the literature. *Disasters*, 23(2), 156–173.

Fothergill, A., & Peek, L. (2004). Poverty and disasters in the United States: A review of recent sociological findings. *Natural Hazards*, 32, 89–110.

Gardner, R. O. (2013). The emergent organization: Improvisation and order in Gulf Coast disaster relief. *Symbolic Interaction*, 36(3), 237–260.

Green, D. (2008). *From Poverty to Power: How Active Citizens and Effective States Can Change the World*. Oxford, UK: Oxfam International.

Gustafsson, M. T., & Johannesson, L. (Eds.). (2016). *Introduktion till politisk etnografi: metoder för statsvetare* [An introduction to political ethnography: Methods for political scientists]. Malmö: Gleerups Utbildning AB.

Hannigan, J. (2012). *Disasters without Borders*. Cambridge, UK: Polity Press.

Holvino, E. (2008). Intersections: The simultaneity of race, gender and class in organization studies [Special issue]. *Gender, Work and Organization*, 17(3), 248–277.

hooks, b. (2010). *Teaching Critical Thinking: Practical Wisdom*. New York: Routledge.

Houten, D. van, & Jacobs, G. (2005). The empowerment of marginals: Strategic paradoxes. *Disability & Society*, 20(6), 641–654.

Jones, E. C., & Murphy, A. D. (2009). *The Political Economy of Hazards and Disasters*. New York, NY: Altamira Press, Rowman & Littlefield Publishers, Inc.

Jordan-Zachery, J. (2007). Am I a black woman or a woman who is black? A few thoughts on the meaning of intersectionality. *Politics & Gender*, 2, 254–263.

Jung, C. (2003). Breaking the cycle: Producing trust out of thin air and resentment. *Social Movement Studies*, 2(2), 147–175.

Kelman, I. (2020). *Disaster by Choice: How Our Actions Turn National Hazards into Catastrophes*. Oxford: Oxford University Press.

Kruks, S. (2001). *Retrieving Experience: Subjectivity and Recognition in Feminist Politics*. Ithaca: Cornell University Press.

Mahrouse, G. (2014). *Conflicted Commitments: Race, Privilege, and Power in Solidarity Activism*. Montreal: McGill Queen's University Press.

McEntire, D. A. (1997). Reflecting on the weaknesses of the international community during the IDNDR: Some implications for research and its application. *Disaster Prevention and Management*, 6(4), 221–233.

Medwinter, S. D. (2021). Reproducing poverty and inequality in disaster: Race, class, social capital, NGOs, and urban space in New York City after Superstorm Sandy. *Environmental Sociology*, 7(1), 1–11.

Neal, D., & Phillips, B. D. (1995). Effective emergency management: Reconsidering the bureaucratic approach. *Disasters*, 19(4), 327–337.

Oliver-Smith, A., & Hoffman, S. M. (1999). *The Angry Earth: Disaster in Anthropological Perspective*. London: Routledge, Taylor & Francis Group.

Parashar, S. (2021, November 4). *Inverting the Gaze: What Wars Know About IR*. Public lecture, Swedish Defence University.

Peek, L. (2019, December 12). The Vulnerability Bearers, Natural Hazards Center [Director's Corner]. Retrieved from https://hazards.colorado.edu/news/director/the-vulnerability-bearers?utm_source=NHC+Master+List&utm_campaign=764c73db6d-EMAIL_CAMPAIGN_2019_01_31_09_35_COPY_01&utm_medium=email&utm_term=0_dabc309806-764c73db6d-54438293

Pelling, M., & Dill, K. (2010). Disaster politics: Tipping points for change in the adaptation of sociopolitical regimes. *Progress in Human Geography*, 34(1), 21–37.

Phillips, A. (1995). *The Politics of Presence*. Oxford: Oxford University Press.

Pilisuk, M., McAllister, J., & Rothman, J. (1996). Coming together for action: The challenge of contemporary grassroots community organizing. *Journal of Social Issues*, 52(1), 15–37.

Reinecke, J. (2018). Social movements and prefigurative organizing: Confronting entrenched inequalities in occupy London. *Organization Studies*, 39(9), 1299–1321.

Schneider, S. K. (1992). Governmental response to disasters: The conflict between bureaucratic procedures and emergent norms. *Public Administration Review*, 52(2), 135–145.

Scott, K., & Liew, T. (2012). Social networking as development tool: A critical reflection. *Urban Studies*, 49(12), 2751–2767.

Seana, S., & Fothergill, A. (2009). 9/11 volunteerism: A pathway to personal healing and community engagement. *Social Science Journal*, 46, 29–46.

Solnit, R. (2009). *A Paradise built in hell: The extraordinary communities that arise in disaster*. London, UK: Penguin Books.

Solnit, R. (2020, March 31). *What Disasters Reveal about Hope and Humanity* [interview by H. McQuilkin and M. Chakrabarti]. WBUR On point.

Retrieved from www.wbur.org/onpoint/2020/03/31/rebecca-solnit-hope-coro navirus

Stallings, R. A., & Quarantelli, E. L. (1985). Emergent citizen groups and emergency management. *Public Administration Review*, 45, 93–100.

Stallings, R. A., & Quarantelli, E. L. (2015). Emergent citizen groups and emergency management. In N. C. Roberts (Ed.), *The Age of Direct Citizen Participation* (pp. 71–103). London, UK: Routledge.

Tierney, K. (2014). *The Social Roots of Risk: Producing Disasters, Promoting Resilience*. Stanford: Stanford University Press.

Thomas, D. S. K., Phillips, B., Lovekamp, D., William, E., & Fothergill, A. (2013). *Social Vulnerability to Disasters* (2nd ed.). London, UK: CRC Press, Taylor & Francis Group.

Twigg, J., & Mosel, I. (2017). Emergent groups and spontaneous volunteers in urban disaster response. *Environment & Urbanization*, 29(2), 443–458.

Voorhees, W. R. (2008). New Yorkers respond to the World Trade Center attack: An anatomy of an emergent volunteer organization. *Journal of Contingencies and Crisis Management*, 16(1), 3–13.

Whittaker, J., McLennan, B., & Handmer, J. (2015). A review of informal volunteerism in emergencies and disasters: Definition, opportunities and challenges. *International Journal of Disaster Risk Reduction*, 13, 358–368.

Wisner, B., Blaikie, P., Cannon, T., & Davis, I. (2004). *At Risk: Natural Hazards, People's Vulnerability and Disasters*. London, UK: Routledge.

Wolford, W. (2006). The difference ethnography can make: Understanding social mobilization and development in the Brazilian Northeast [Special issue]. *Qualitative Sociology*, 29(3), 335–352.

Young, I. M. (2000). *Inclusion and Democracy*. Oxford: Oxford University Press.

# 2 Most Affected, Least Heard

## Most Affected...

Add short-term chocks to long-term marginalization, shake it, and you have a perfect storm of vulnerability. Long before Sandy, Rockaway's low-income communities of color faced problems of unemployment, lack of education, geographic isolation, inadequate transportation, and shortage of essential services. When Hurricane Sandy hit the peninsula in October 2012 – and did so with tremendous force – these underlying problems intertwined with the acute mess of an area instantly rattled to its core. A year later, on November 11, 2013, sociologist Max Liboiron made a presentation at the Superstorm Research Lab.

> It's not over. The beginning stretched to way before the storm. The storm was a punctuation mark at a much larger crisis. These places did not have clean water before this. These places did not have access to health care. There is no over. There is no restoration to a place that was already ground zero before the storm even hit.
>
> (Liboiron, 2013)

Vulnerability has been defined as the "conditions created by physical, social, economic, and environmental factors or processes, which increase the susceptibility of a community to the impacts of hazards" (Thomas et al., 2013, p. 42). Social vulnerability, to specify further, are those characteristics of a person or group that affect their capacity to predict, withstand, and recover from a disaster (Thomas et al., 2013). The framework of vulnerability developed by Ben Wisner, J. C. Gaillard, and Ilan Kelman (2012) captures the extent to which social status determines if someone is impacted by a natural hazard, as well as the processes that led to and preserve these social statuses.

DOI: 10.4324/9781003005278-2

When thinking about vulnerabilities of low-income communities of color, words matter. As Elizabeth Marino and A. J. Faas (2020) state, the everyday use of the word "vulnerable" can act to "flatten and simplify diverse communities" (p. 33). Labeling particular groups as vulnerable might in itself be an act of marginalization. Focus tends to be on characteristics of the group, rather than on characteristics of the social inequalities that distribute risk and vulnerability. In response to this critique, this chapter outlines a number of structural factors deeply woven into the material fabric of society and illustrates how they result in differentiated vulnerability for socioeconomically marginalized communities of color.

Some social groups systematically lose from the social order when it comes to the effects of disasters and their possibilities to influence these circumstances. As Kelman (2020) states when comparing the effects of Hurricane Matthew on the multimillion dollar mansions of Hilton Head, South Carolina, USA, with the ramshackle housing climbing up the slopes of Mexico City's Popocatepetl volcano, the "rich population of South Carolina can purchase insurance, can self-evacuate in private vehicles, can afford to take several days off work," whereas the poor communities of Mexico City lack resources and opportunities for improving their situation (p. 46). With regard to Hurricane Katrina, the systematic neglect by government agencies that put low-income communities of color at higher risks than white high-income groups – and led to a higher death toll for poor and minority communities – is an illustrative case (Elliot & Pais, 2006; Price, 2008; Stivers, 2007). Everyone in New Orleans knew that the hurricane was coming; yet, as Bauman notes in his 2011 book *Collateral Damage*, not everyone could

> scrape together enough money for flight tickets. They could pack their families into trucks, but where could they drive them? Motels also cost money, and money they most certainly did not have. And – paradoxically – it was easier for their well-off neighbors to obey the appeals to leave their homes, to abandon their property to salvage their lives: the belongings of the well-off were insured, and so Katrina might be a mortal threat to their lives, but not to their wealth.
>
> (Bauman, 2011, p. 6)

Bauman's observations were substantiated by a study by Deng et al. (2021) showing how wealth and race were associated with differences in evacuation patterns, where low-income communities of color were less able to relocate. Bauman talks about "collateral" casualties, as they were seen as unimportant or came as surprise effects of an urban planning process that did not take matters of inequality into account (Bauman, 2011).

Issues of residence status also play a role, especially in the United States, where many undocumented immigrants shy away from recovery assistance for fear of deportation (Fothergill et al., 1999). Oftentimes, low-income communities live in neighborhoods near transportation routes or industrial corridors as a result of historical patterns of enforced segregation and discriminatory zoning regulations (Tierney, 2014). As the environmental justice literature have demonstrated, ethnic minorities, people of color, indigenous communities, and low-income communities face a higher burden of environmental exposure from air, water, and soil pollution from industries. Race and class intersect with the siting of toxic waste, leaving low-income people of color disproportionally affected by environmental hazards (Bullard et al., 2008; Mohai et al., 2009).

Lack of safe and affordable housing is another issue that often puts low-income communities of color at higher risk (Adams et al., 2009; Collins, 2005; Dooling & Simon, 2012; Wisner et al., 2004). In the United States, housing access is often interlinked with issues of discrimination based on race, gender, health/ability, and age (Mooser, 1998; Pelling, 2003; Sandersson, 2000; Thomas et al., 2009). Research also shows how racial prejudice inherent in city planning disproportionally place minority and immigrant population at risk (Dooling & Simon, 2012). Discriminatory policies limit housing options for poor people of color, which confines them to neighborhoods that are unpopular among more resourceful people (Bullard, 1993; Fothergill et al., 1999). Thus, as part of ongoing trends of race and class-based gentrification in the United States, socioeconomically marginalized people of color are warehoused away from services and jobs.

In the areas available to them, investment is hard to attract since the value of property is declining. Economic downturns are often felt harder in these areas since layoff policies often disadvantage black and Latino people, sometimes resulting in an outflux of both local business and larger employers. Moreover, politicians are generally more responsive to neighborhoods that are populated by affluent and (oftentimes) white people, where schooling, policing, fire protection, garbage removal, and other social services are higher prioritized than in lower-class neighborhoods. Poorly maintained infrastructure and housing often result in the isolation of neighborhoods. As a result, many black and Latino neighborhoods are populated by people who are poorly educated, whose prospects for employment are bad, and who live around a higher concentration of crime (Gupta, 2013; Klein, 2012; Young, 2000).

These ongoing housing trends affect what happens in disaster situations, as low-income communities of color tend to face housing problems after a disaster strikes their area. Challenges include living in unsafe buildings with greater exposure to disasters, as well as having to deal with housing shortages following a disaster. The Loma Prieta earthquake in Northern California, for example, mostly displaced the elderly, the homeless, and low-income Latinos. After Hurricane Hugo in South Carolina, out of the 60,000 people who became homeless, most were of low-income and ethnic minorities (Fothergill & Peek, 2004). Moreover, disaster rebuilding and recovery figure as causes of displacement for low-income groups (both homeowners and renters). Low-income homeowners often prefer to sell their homes rather than take on the extra cost of rebuilding disaster-damaged property, as they often cannot afford to rebuild according to federal and national regulations (Whittle, 2005). When considering building codes, for instance, middle- and high-income households will be more likely to have the necessary resources to elevate their homes, a requirement for some insurance options (Gupta, 2013). Racial differences in insurance settlement claims were found after Hurricane Andrew, where black neighborhoods were less likely to have insurance with major companies due to redlining practices (Peacock et al., 1997).

Low-income renters are often even worse off. In post-disaster repair, services are geared toward homeowners and legal tenants, excluding multifamily and affordable housing units. Some studies show that low-income renters are the least likely of all households to receive emergency assistance in terms of repairs (Fothergill & Peek, 2004). After the Loma Prieta earthquake, single-family homes were rebuilt at a much faster pace than multifamily units – occupied by low and moderate-income renters – which remained unrepaired for many years following the disaster (Comerio et al., 1994). Among low-income renters, public housing tenants are often affected most severely. After Hurricane Katrina, public housing tenants were evacuated to the Federal Emergency Management Agency's (FEMA) trailer parks outside of the city and were ineligible for much of the aid homeowners could apply for. Available affordable housing then dropped, since many houses were destroyed and rents soared to levels that were out of range for previous tenants, preventing their return. Furthermore, the city implemented plans to tear down storm-affected public housing apartment buildings and in their place make room for mix-income rental units (Adams et al., 2009). Affordable housing, which there is a shortage of in non-disaster situations, can become even scarcer as reconstruction demands require landowners to raise the rents in order to afford the rebuilding (Gupta, 2013). After the Whittier Narrows earthquake in 1987, many low-income tenants were evicted for late rent payment, although the earthquake had occurred on the same day as rent was due, preventing many of the tenants to pay it (Fothergill & Peek, 2004). Further, it is not uncommon for higher-income evacuees to obtain surplus housing in a community, while low-income communities face a problem finding rental housing after disasters (Fothergill & Peek, 2004).

### ...and Least Heard

On August 6, 1965, 51 years before Hurricane Sandy wreaked havoc on the American East Coast, President Johnson signed the Voting Rights Act to stop the use of literacy tests as voting requirements. Prior to the signing, voting activists in the Selma-to-Montgomery-March had been sprayed with tear gas and beaten bloody by Alabama state troopers (PennState University Libraries, n.d.). As a result of age-old processes of formal and informal discrimination and racism in the United States, political participation – in other words everything from voting to getting engaged in civic work and activism – is lower among low-income communities of color as compared to participation among more privileged social groups (Frampton et al., 2008; Stoll & Wong, 2007). The legacy of racism with regard to political marginalization dates back to the era of slavery, where enslaved persons were exempted from the right to vote. Once African Americans gained the formal right to vote, they were threatened and beaten by white supremacists when they attempted to exercise that right. Literacy tests were merely yet another measure that functioned to the disadvantage of poor people of color.

Today's disenfranchisement regulations (being deprived the right to vote if you have been incarcerated) (Alexander, 2012; Tucker, 2009) disproportionally diminish voting rights for people of color – primarily African Americans – the group that is the primary target of racialized mass incarceration (Forman, 2010; Frampton et al., 2008; Haney López, 2010). Explicitly discriminatory regulations notwithstanding there are also more subtle forms of political marginalization. Distrust toward the political system is high among many low-income communities of color. Such distrust is often for good reasons. Many political institutions have either not taken these communities' interests into account or have actively worked against their well-being (Griffin & Newman, 2008). However, even among those who are actively seeking engagement, there are other types of barriers to political participation such as language barriers or lack of equal education opportunities that affect the level of self-confidence or knowledge needed to partake on equal terms (Tucker, 2009). Finally, it could also be a matter of finding time and resources for political participation. Many low-income persons of color spend large chunks of the day working in order to survive. As of 2015, over 800,000 New Yorkers were officially below the federal poverty line while still employed, as their wages did not meet basic needs (Fiscal Policy Institute, 2015).

In sum, low-income communities of color face ongoing socioeconomic marginalization and are often put in harm's way of disasters as a result. Such marginalization often goes hand in hand with lack of political voice in civic and political work. Yet, risks are produced by decisions that communities, societies, organizations, and political actors make (Tierney, 2014). Within processes of urban planning, for example, differential exposures to disasters across social groups are often the outcomes of local political economies where control in decision-making with regard to land use and development mainly rests with elites (Tierney, 2014). Therefore, in the wake of disasters, the most affected are also the politically less heard.

## Conclusion

This chapter has shown how low-income communities of color are exposed to political processes that put them in harm's way of oppressive conditions, be it structural and historical patterns of discrimination, or the sudden on-set perils of disasters. It has also demonstrated how the two create unequal disaster vulnerability. It is however important to stay mindful not to discursively nullify the agency of people belonging to this social category (Marino & Faas, 2020). It is, as Judith Butler, Zeynep Gambetti, and Leticia Sabsay (2016) put it, always risky to state that "socially disadvantaged groups are especially vulnerable" (p. 2). It may obscure the way that every social group is in itself constituted by intersecting identities, as well as hide the fact that the delineations of social groups are always under discursive contestation (Butler, 1990). It essentializes identities into locked categories such as "woman," "Black," "Latino," or "disabled." This simplifies complex identity structures and renders invisible liminal and intersectional subject positions.

In exploring the story of Occupy Sandy in Rockaway, I talked to activists, but perhaps more importantly, I listened to and learned from vulnerable and marginalized residents. What they taught me speaks volumes about resistance, dignity, and empowerment in the face of disasters. In showing how, in the transition from relief to recovery, rosy collaboration turned into contentious micro-politics, I aim to illustrate the nuances, complexities, and multi-faceted variations of what vulnerability, identity, and empowerment means in a post-disaster social justice space. Chapter 3 will delve deeper into the issue of social change and empowerment in the aftermath of disasters.

## References

Adams, V., Van Hattum, T., & English, D. (2009). Chronic disaster syndrome: Displacement, disaster capitalism, and the eviction of the poor from New Orleans. *American Ethnologist*, 26(4), 615–636.

Alexander, M. (2012). *The New Jim Crow: Mass Incarceration in the Age of Colorblindness*. New York: The New Press.

Bauman, Z. (2011). *Collateral Damage: Social Inequalities in a Global Age*. Cambridge, UK: Polity Press.

Bullard, R. (Ed.). (1993). *Confronting Environmental Racism: Voices from the Grassroots*. Cambridge, MA: South End Press.

Bullard, R. D., Mohai, P., Saha, R., & Wright, B. (2008). Toxic Wastes and Race at Twenty: Why Race Still Matters After All of These Years. *Environmental Law*, 5(29), 371–411.

Butler, J. (1990). *Gender Trouble*. New York: Routledge.

Butler, J., Gambetti, Z., & Sabsay, L. (2016). *Vulnerability in Resistance*. Durham, NC: Duke University Press.

Collins, T. (2005). Households, forests and fire hazard vulnerability in the American West: A case study of a California community. *Global Environmental Change: Environmental Hazards*, 6(1), 23–27.

Comerio, M. C., Landis, J. D., & Rofe, Y. (1994). *Post-disaster Residential Rebuilding: Working Paper 608*. Berkeley, CA: Institute of Urban and Regional Development, University of California.

Deng, H., Aldrich, D. P., Danziger, M. M., Gao, J., Phillips, N. E., Cornelius, S. P., & Wang, Q. R. (2021). High-resolution human mobility data reveal race and wealth disparities in disaster evacuation patterns. *Humanities and Social Sciences Communications*, 8(1), 1–8.

Dooling, S., & Simon, G. (2012). *Cities, Nature and Development: The Politics and Production of Urban Vulnerabilities*. Farnham, Surrey, UK: Ashgate Publishing Ltd.

Elliot, J. R., & Pais, J. (2006). Race, class, and Hurricane Katrina: Social differences in human responses to disaster. *Social Science Research*, 35(2), 295–321.

Fiscal Policy Institute. (2015, October 8). *FPI-NELP Response to NYS Business Council Statement on Gov. Cuomo's $15 Minimum Wage Proposal*. Retrieved from http://fiscalpolicy.org/fpi-nelp-response-to-nys-business-council-statement-on-gov-cuomos-15-minimum-wage-proposal

Forman, J., Jr. (2010). Why care about mass incarceration. *Michigan Law Review*, 108(6), 993–1010.

Fothergill, A., & Peek, L. (2004). Poverty and disasters in the United States: A review of recent sociological findings. *Natural Hazards*, 32, 89–110.

Fothergill, A., Maestas, E. G. M., & DeRouen Darlington, J. (1999). Race, ethnicity and disasters in the United States: A review of the literature. *Disasters*, 23(2), 156–173.

Frampton, M. L., Lopez, I. H., & Simon, J. (2008). *After the War on Crime: Race, Democracy and New Reconstruction*. New York, NY: New York University Press.

Gupta, A. (2013, January 28). Disaster Capitalism Hits New York; the city will adapt to flooding – but at the expense of the poor? *In These Times*. Retrieved from http://inthesetimes.com/article/14430/disaster_capitalism_hi ts_new_york

Griffin, J. D., & Newman, B. (2008). *Minority Report, Evaluating Political Equality in America*. Chicago, IL: University of Chicago Press.

Haney López, I. F. (2010). Post-racial racism, racial stratification and mass incarceration in the age of Obama. *California Law Review*, 98(3), 1023–1074.

Kelman, I. (2020). *Disaster by Choice: How Our Actions Turn National Hazards into Catastrophes*. Oxford: Oxford University Press.

Klein, N. (2012, November 6). Hurricane Sandy: Beware of America's Disaster Capitalists. *The Guardian*.

Liboiron, M. (2013). Turning the Tide: Remember Sandy, Revive Our City. Procession and Rally! *Superstorm Research Lab*. Retrieved from http://superstormresearchlab.org/2013/07/27/turning-the-tide-remember-sandy-revive-our-city-procession-and-rally/

Marino, E. K., & Faas, A. J. (2020). Is vulnerability an outdated concept? After subjects and spaces, *Annals of Anthropological Practice*, 44(1), 33–46.

Mohai, P., Pellow, D., & Roberts. T. (2009). Environmental justice. *Annual Review of Environment and Resources*, 34 (1), 405–430.

Mooser, C. (1998). The asset vulnerability framework: Reassessing urban poverty reduction strategies. *World Development*, 26(1), 1–9.

Peacock, W. G., Gladwin, H., & Morrow, B. H. (1997). *Hurricane Andrew: Ethnicity, Gender and the Sociology of Disasters*. New York, NY: Routledge.

Pelling, M. (2003). Toward a political ecology of urban environmental risk. In K. S. Zimmer and T. J Basket (Eds.), *Political Ecology: An Integrative Approach to Geography and Environment-Development Studies* (pp. 73–93). New York, NY: Guildford Publications.

PennState University Libraries. (n.d.). *Selma-to-Montgomery March*. Retrieved from: https://libraries.psu.edu/about/collections/jack-rabin-collection-alab ama-civil-rights-and-southern-activists/alabama-civil-4

Price, G. N. (2008). Hurricane Katrina: Was there a political economy of death? *Review of Black Political Economy*, 35(4), 163–180.

Sandersson, D. (2000). Cities, disasters and livelihoods. *Environment and Urbanization*, 12(2), 93–102.

Stivers, C. (2007). "So poor and so black": Hurricane Katrina, public administration, and the issue of race. *Public Administration Review*, 67(1), 48–56.

Stoll, M. A., & Wong, J. S. (2007). Immigration and civic participation in a multiracial and multiethnic context. *The International Migration Review*, 41(4), 880–908.

Thomas, D. S. K., Phillips, B. Fothergill, A., & Blinn-Plike, L. (2009). *Social Vulnerability to Disasters*. London, UK: Taylor and Francis.

Thomas, D. S. K., Phillips, B., Lovekamp, D., William, E., & Fothergill, A. (2013). *Social Vulnerability to Disasters* (2nd ed.). London, UK: CRC Press, Taylor & Francis Group.

Tierney, K. (2014). *The Social Roots of Risk: Producing Disasters, Promoting Resilience*. Stanford, CA: Stanford University Press.

Tucker, J. T. (2009). *The Battle Over Bilingual Ballots, Language Minorities and Political Access Under the Voting Rights Act*. Farnham, UK: Ashgate Publishing Limited.

Whittle, P. (2005, September 4). Gentrification through natural disaster: Old neighborhoods around Edgewater Drive and the county are changing because of Charley. *Sarasota Herald Tribune*.

Wisner, B., Blaikie, P., Cannon, T., & Davis, I. (2004). *At Risk: Natural Hazards, People's Vulnerability and Disasters*. London, UK: Routledge.

Wisner, B., Gaillard, J. C., & Kelman, I. (2012). *The Routledge Handbook of Hazards and Disaster Risk Reduction*. New York, NY: Routledge.

Young, I. M. (2000). *Inclusion and Democracy*. Oxford: Oxford University Press.

# 3 Ain't No Power Like the Power of the People

## Post-disaster Empowerment Processes

The first section of this chapter elaborates on how empowerment is theorized within emancipatory studies and in literature on emergent groups. In wedding studies on emergent groups to broader emancipatory theories, I conceptualize empowerment of marginalized communities in post-disaster processes as twofold: taking active part in relief work (as opposed to passively receiving assistance) and being able to influence both ends and means of the work.

The emancipatory literature has understood empowerment as a process in which people gain increasing control over their own lives through participation in self-organized arenas of social change (Chambers, 1983, 1997; Chavis, 2001; Cornwall, 2002, 2016; Dacombe, 2018; Fung & Wright, 2003; Hilmer, 2010; Houten & Jacobs, 2005; Jung, 2003; Pateman, 1970; Pilisuk et al., 1996; Scott & Liew, 2012). Originally, the concept described grassroots struggles to transform unequal power relations, closely linked with consciousness-raising. Today, the concept of empowerment spans academic boundaries as well as travels across theory and practice, be it in the field of international development or in social justice movements. Consciousness-raising, elaborated by bell hooks (1994) and associated with Paulo Freire's "critical pedagogy" (2005), is a process in which marginalized people come to understand their own social position and how it stands in relation to power dynamics, such as racism, sexism, or class elitism (Batliwala 1994; hooks, 1994; Young 2000). As Srilatha Batliwala avers (1994), the process of empowerment has a transformational potential since it leads people to start seeing themselves as entitled to make decisions over political issues that affect them.

Within international development practice, empowerment has become one of the most elastic buzzwords (Cornwall, 2016). The flourishing of the concept came about as part of the "participatory turn"

DOI: 10.4324/9781003005278-3

(Cooke & Kothari, 2001), where western-biased top-down approaches were challenged by radical empowerment discourses rooted in Freirean pedagogy (Freire, 2005). The participatory turn advocated that marginalized groups take part in decision-making, from the bottom up (Chambers, 1983, 1997). This is similar to what Eliosoph (2011) refers to as "empowerment talk": a rhetoric that stresses open, egalitarian, and voluntary membership, and transparent, unbureaucratic practices.

The idea of empowerment also cuts to the core of many social justice movements that develop critical perspectives on the social order (Choudry et al., 2012). Social justice movements have a capacity to push the agenda for what should be considered a political problem. They may mobilize marginalized communities and bring forth neglected problem formulations and, in so doing, politicize issues traditionally considered to fall outside of the realm of institutional politics. We witnessed this with the civil rights movement, which turned the US's deeply institutionalized racism into a political problem, as well as with feminist movements across the globe that have put issues of domestic violence against women on political agendas. We have seen the contemporary Black Lives Matter movement expose institutionalized racism within US police forces (Black Lives Matter, 2016), and indigenous groups joining forces with environmental activists to challenge environmental destruction (D'Angelo & O'Connor, 2016).

As Cornwall (2016) notes, empowerment is a process, not an end goal or a fixed state. Empowering marginalized groups is important regardless of whether it leads to any objective change of the social and economic structures, as discussed as well by Catherine Campbell (2014). To reach actual transformation, practices of decision-making are just as important, according to Iris Marion Young (in Cudd & Andreasen, 2005). Focus thus is on improved capacities for change (rather than change itself). Empowerment is about whom gets to decide what to do, and how it is to be done. This is best explored by paying close attention to micro-political processes, organizational forms, and techniques of decision-making.

Participants are more likely to experience empowerment when "practices and activities tend to be fluid, tasks, skills and resources tend to be shared and decision-making and leadership is collectively shared" (Choudry et al., 2012, p. 159). Reinecke (2018) names it prefigurative organizing, meaning that social inequalities are directly transformed in the organizing process by way of shifting power relations on a micro-political level while simultaneously working toward the end goal of wider structural change. In such organizing, ends and means are aligned, and everyday practices bring alternative ways of relating to others. In this way, empowerment processes are situated in larger oppositional struggles against structural inequality. Campbell writes about an emerging net "of small-scale acts of resistance to inequality, pockets of social protests apparently randomly blossoming in local contexts all over the world" (Campbell, 2014, p. 53). Small-scale projects of activism, often improvisational in character, are part of a global movement toward transformation of inequality (Campbell, 2014).

Summing up the emancipatory literature on empowerment, I understand empowerment of marginalized groups to be a process, best captured by exploring organizing techniques that aim to alter power dynamics and let agenda setting be in the hands of those who are to be empowered. It is also important whether marginalized people gain influence over how the work is organized, since those who have the freedom to shape a forum are the ones with the most power (Cornwall & Schatten Coelho, 2007; Hickey & Mohan, 2004). How then should we understand empowerment in the particular setting of a post-disaster process? Some preliminary answers can be found in the literature on emergent groups, to which I now turn.

## Emergent Groups in Post-disaster Settings

As will be further discussed below, a few studies have paid attention to the role of emergent groups in empowering marginalized disaster-affected communities (see Carlton, 2015; Seana & Fothergill, 2009; Solnit, 2009; Twigg & Mosel, 2017). Most of the literature on emergent groups however does not focus on empowerment at all. Instead, focus is on emergent groups' organizing mode, the effectiveness of their operations, or their relation to more formal authorities. Research on emergence – first explored in the 1950s by sociologists focusing on collective behavior – focuses on how individuals, organizations, or networks respond in new and improvisational ways, confronted as they are with "situations and problems that lie outside the bounds of normal, everyday existence" (Schneider, 1992, p. 137).

Emergence stands in opposition to a command-and-control approach that promotes strict, rigid, and bureaucratic norms with centralized, hierarchical decision-making (Bondesson, 2020; Drabek & McEntire, 2003; Dynes, 1994; Gardner, 2013; Harrald, 2006; Helsloot & Ruitenberg, 2004; Neal & Phillips, 1995; Schneider, 1992; Strandh & Eklund, 2018; Turner & Killian, 1972; Whittaker et al., 2015). Hence, emergent responses often fall outside of regular behavior patterns. As Russel. R. Dynes (1994) notes, "traditional roles are expanded. New roles are created. Organizations are transformed. New actors, both individual and collective, assume new responsibilities" (p. 2). Emergent groups lack formal structures for group membership or roles and have no predesigned tasks or rules (Drabek & McEntire, 2003; Dynes, 1970, 1987; Gardner, 2013; Stallings & Quarantelli, 1985, 2015; Voorhees, 2008). They operate in the fast-changing and volatile environments of emergencies where new information about needs constantly arrive, sometimes in haphazard manners (Drabek & McEntire, 2003).

Emergent groups often come together as a response to emergencies where formal authorities are perceived to be failing (Bondesson, 2020; Stallings & Quarantelli, 1985). Failures to assist marginalized communities in particular have since long been noted in disaster research (Aptekar, 1990). After Hurricane Hugo, many affected people faced poverty, illiteracy, and physical isolation and were living out of sight of public authorities, in unmarked homes or on unmapped roads, which meant that they received little or no aid (Fothergill & Peek, 2004). Sometimes, more blatant racial prioritizing happen, with responses targeted to white areas before reaching socioeconomically marginalized ones (Beady & Bolin, 1986). Language can also be an issue (Fothergill et al., 1999). After Hurricane Andrew, women of color who spoke no English became targets of dishonest practices of construction contractors (Peacock et al., 1997). Immigrant communities sometimes avoid public officials and relief workers because of previous experiences of political repression, or precarious immigration statuses (Peacock et al., 1997; Phillips, 1993).

Doubt in public authorities' capacity to deal effectively or justly with the situation can mobilize people to offer spontaneous help. For example, a variety of new, nontraditional behavior surfaced after Hurricane Katrina in Louisiana (Gardner, 2013; Rodríguez et al., 2006). Emmanuel (2006) shows how the misguided targeting of people of color by law enforcement in New Orleans became a trigger for emerging social justice mobilization. Emergent groups not only provided relief but also resisted both stereotypical and racist media imagery and skewed relief from city and federal agencies. Gardner (2013) describes various emergent groups in the aftermath of Hurricane Katrina and Hurricane Ike that responded to the vacuum left by established relief organizations. Theirs was a "decentralized and radically democratic organizational model that privileged spontaneity over bureaucracy and valued unscripted and improvised behaviors" (Gardner, 2013, p. 245). Incoming volunteers were valued for whatever competence they brought. Volunteers questioned centralized, hierarchical leadership; they allowed room for improvisation and cultivated intentional norms of mutual respect, nonexclusion, and consensus (Gardner, 2013).

In his examination of an emergent group which formed in the aftermath of the 9/11 attack in NYC, Voorhees (2008) characterized it by its lack of hierarchical leadership, lack of written rules, and fleeting membership. The Student Volunteer Army in Christchurch, New Zealand, emerged to assist in cleaning up liquefaction after the 2010–2011 earthquake sequence, and Blaze Aid was a group formed by farmers after the 2009 bushfire in Victoria, Australia (Whittaker et al., 2015). After the bushfire in Southern California in 1970, volunteers developed an organized effort to handle registrations of evacuees, and after the Loma Prieta earthquake, there were ad hoc efforts coordinating the response of volunteer organizations (Neal & Phillips, 1995).

Emergent groups could be said to operate according to an anarchist principle of autonomy. Autonomous organizing has historically been understood as an integral part of anarchist organizing, with focus on activities that can emerge when people come together and interact in an open-ended manner (Dixon, 2012). Autonomy is believed to increase marginalized people's influence and allow them to take active part in problem definition and implementation of solutions (Holvino, 2008; Pilisuk et al., 1996; Yeich, 1996). Another notion that underpins anarchist principles of autonomy is that formally structured organizations put unnecessary strain on spontaneity. Fluid and malleable processes are preferred, as such processes are more responsive to democratic impulses (Smith & Glidden, 2012). Linear planning is thus refuted in favor of improvisation, shared learning, and organic action (Campbell, 2014).

With regard to empowerment, Steffen Seana and Alice Fothergill (2009) found that volunteering following the disaster event had a meaningful therapeutic effect, as it prevented feelings of victimization. Volunteering impacted long-term personal healing and fostered new levels of identification with, and affinity for, other community members. Sally Carlton (2015) demonstrated that volunteering led to a sense of well-being among refugee youth involved in the Student Volunteer Army in Canterbury, New Zealand, following the earthquakes in 2010–2011. Marginalized on both ethnicity and age basis, the youths experienced a heightened sense of belonging and developed leadership skills. In volatile emergency situations, marginalized groups may get the chance to step up and take on roles previously inaccessible to them. Such exercise of agency has empowering effects. Instead of being a victim in need of help, taking action increases a person's sense of self and belonging and helps shift their identity "from passive victim to active engager" (Carlton, 2015, p. 344).

Solnit's (2009) discussion of the empowerment that emerged after the Mexico City earthquake is another example of the positive effects of engaging in disaster relief. Dynes (1987), leaning against socio-logical role theory, argues that disaster victims are socially constructed through a dialectic process between those affected by a disaster and those coming to help. Victims and helpers thus co-constitute each other because "it is not possible to be a helper without any victims" (Dynes, 1987, p. 89). To be a helper is a much more socially rewarding role than to be a victim. Based on this theorizing, I propose that taking an active part in the work, as opposed to passively receiving assistance, is key for empowerment in disaster relief. In other words, to step out of the socially constructed role of victim and into the helper role.

**Inside Out or Outside In: A Puzzle of Empowerment**

Social justice spaces are often run by socially, economically, and polit-ically privileged activists who work with marginalized communities to empower them to transform unequal social relations (Baiocchi et al., 2014; Campbell, 2014). As Campbell (2014) avers, "a generation of activists has defined its role as working with marginalized communi-ties to develop their collective agency to resist and transform unequal social relations" (p. 47). Yet at the same time, an idea that mark both practice and theory is that genuine empowerment comes from within the marginalized community, in self-organized forums. Empowerment is not something that can be "bestowed by others, it cannot be done to or for anyone else," as Cornwall states (2016, p. 343). Despite the belief that genuine empowerment comes from within, the start-up and overall management of emancipatory projects are most often done by actors who do not themselves belong to marginalized communities (Campbell, 2014; Cornwall, 2003; McDaniel, 2002; Pilisuk et al., 1996; Snow et al., 2004). This is due to social stratification that distributes cultural and political capital unevenly. Privileged groups of people often exhibit stronger capacities to engage in social justice movements than nonprivileged groups (Juris et al., 2012, p. 3436; Snow et al., 2004, p. 117).

This mismatch between theory and practice creates tensions. The trouble that arises when participants in social justice spaces occupy different positions in social hierarchies has been thoroughly discussed by democratic theorists, feminist theorists, and social movement scholars (Fung, 2005; Karpowitz et al., 2009; Mansbridge et al., 2010; Pateman, 1970; Young, 2001). In societies structured by inequalities, privileged people may inadvertently use their social, political, and educational capital to shape agendas of social justice spaces (Baiocchi et al., 2014; Campbell, 2014; Cornwall, 2003; Juris et al., 2012; Mahrouse, 2014;

Pilisuk et al., 1996; Snow et al., 2004). Angela Davis (2016) claims that one of the problems is that privileged people think of nonprivileged people as receivers of charity instead of equal partners. Thus, the outlook merely reproduces the unequal relation, as it constitutes nonprivileged people as inferior (Davis, 2016, p. 26). Within discussions on international development, the problem has been branded the "white savior complex" (Cole, 2012). It is "the idea that you, as a single (and possibly unskilled) foreigner, can save a whole community. This sort of savior complex is condescending because it implies that you're a hero while those locals are helpless" (Ferguson, 2016). The savior complex is increasingly ridiculed and critiqued in that donor organizations portray complicated issues of poverty through oversimplified images of helpless children that need saving from foreigners (Randhawa, 2016). The phenomenon underscores struggles within social justice movements as well, for example, within feminist online movements. Here, women of color question white feminists for dominating the sphere, which in turn replicates some of the same inequalities that the movements seek to address (Holm & Castro, 2018; see also RUMMET, 2013, 2014; and hashtag #solidarityisforwhitewomen).

Even if social justice spaces may be explicitly set up to overcome issues of power and privilege, the problem of inequality may surface (Baoicchi et al., 2014; Cornwall & Schatten Coelho, 2007; Davis, 2016; Hickey & Mohan, 2004; Holvino, 2008; Pilisuk et al., 1996). As Mahrouse (2014) discusses, many activists cannot help "but re-enact a colonizing role" (p. 96), such as when solidarity activists traveling to Palestine assume paternalistic attitudes, even if they are fully aware of their privileged position in relation to Palestinians. The "process of coming to terms with, or acknowledging one's racialized power is difficult and fraught with contradiction, and the (re)productive tensions of whiteness that exist within antiracist practices entail ambivalence, doubt, and ethical struggle" (Mahrouse, 2014, p. 95). Social hierarchies might hence be reproduced even in projects where participants are well aware of them and even though these projects may be explicitly geared to overcome inequalities (Cornwall & Goetz, 2005, p. 793; Holvino, 2008, p. 18; Pilisuk et al., 1996, p. 31).

Feminist scholars of democracy have discussed how norms, although seemingly egalitarian, may be skewed. Norms about how interests are put forth, which type of voice is perceived as knowledgeable, or how conflicting interests should be worked out, may be biased (Cornwall & Schatten Coelho, 2007; Hickey & Mohan, 2004; Mansbridge, 1976; Mansbridge et al., 2010; Pateman, 1970; Young, 2001). For example, Jane Mansbridge (1976) explored class differences that resulted in uneven capacities to communicate in a style that generated respect and recognition. Apart from ostensibly universal norms that are in fact skewed, marginalized people may struggle with self-doubt. The self-confidence to make statements, articulate claims, or challenge rules may be linked to a person's position in social hierarchies (Alfrey & Twine, 2017; Correll, 2001; Eagly & Carli, 2007; Hayward, 2004; Holm, 2019; Livingston et al., 2012; Ridgeway & Kricheli-Katz, 2013; Young, 2001). Feminist and postcolonial thinkers have theorized this internalized self-doubt. Sonia Kruks (2001), in reading Frantz Fanon and Simone de Beauvoir, describes how marginalized persons' experience shame related to the objectifying look of privileged people (p. 99). Another concept is minority stress, describing the kind of chronic stress that gay men felt based on experiences of discrimination and harassment, which in turn had resulted in negative self-understandings (Meyer, 1995).

Feminist thinkers have offered some radical suggestions for counteracting inequality. For instance, some degree of differentiation among participants may be necessary, based on identity-markers such as gender, race, able-bodiedness, class, or sexual orientation. Anne Phillips (1995) called it "the politics of presence" in opposition to liberal democracy and its focus on the politics of ideas. Compensatory mechanisms (often termed affirmative action) are designed to acknowledge and offset the problem of inequality (Bacchi, 1996). On a macro-level, this could mean quota systems to achieve gender parity in parliamentary systems or re-writing constituent boundaries to raise the number of elected minority representatives (Phillips, 1996). In micro-political spaces, compensatory mechanisms might be "progressive stacks" whereby marginalized people are allowed longer time in speaking rounds, or demographic restrictions on trainers, facilitators, or leaders to limit the number of privileged people occupying positions of power. Sensitization techniques are also sometimes used. One example is activists that explicitly refer to their own race-, ethnic-, or gender-based privileges as a way of acknowledging and countering their importance (Ahmed, 2004; Baiocchi et al., 2014; Mahrouse, 2014). Micro-political spaces seem to need such compensatory mechanisms in order to counteract domination by privileged people (Campbell, 2014; della Porta, 2005; Dixon, 2012; Dovi, 2009; Fraser, 1990; hooks, 2010; Smith & Glidden, 2012; Young, 2001).

Yet, compensatory mechanisms of differentiation often bring about their own problems. Ahmed (2004) problematizes the notion of proclaiming one's whiteness as a way of offsetting the power that comes with it. If racism per definition is something that goes unchecked, something that the person is unaware of, then drawing attention to one's whiteness becomes a way of absolving oneself of racism. Declaring whiteness becomes an empty speech act that in fact does little to undo the power. Ahmed also point to the fact that whiteness is only invisible to white people. Everyone else is made acutely aware of it, as white bodies constitute the norm in most spheres of life. This means that declaring whiteness functions as a form of white self-centeredness.

Another problem with compensatory mechanisms based on differences is that they risk essentializing social identities, as noted by among others Carol Gould (1996). Phillips (1996) exemplifies this in her discussion of how antiracist activists take issues with stereotypical descriptions of black and white people. Yet, as they struggle for this problem to be acknowledged through strategies that highlight differences, they end up obscuring the cultural and religious pluralism among people of color (Phillips, 1996). The focus by black feminist researchers, such as Kimberly Crenshaw (1991), on intersectional analysis further raises the complexity inherent in issues of community and identity. Groups are never mutually exclusive since various social positions intersect (Young, 2000).

In response to skewed norms and power dynamics hidden behind ostensible empowerment projects, previous researchers have demonstrated forms of subtle resistance that nonprivileged actors have used. For example, James Scott (1985) explored peasants' resistance to elite classes and showed that the peasantry seldom resisted through full-blown explicit protests because that would create overwhelming backlashes. Instead, they resisted through low-intensity expressions, such as subtly ignoring decrees, engaging in boycotts, thefts, quiet strikes, and even malicious gossip – all the while keeping a façade of compliance. Similarly, Mosse (2005) explored how communities in British development projects in India exercised a form of silent agency and shaped the course of the projects in a way that was not immediately observable. Likewise, Campbell (2014) described how efforts by white women to strengthen women's networks in the global south were challenged by black women. They claimed that they had in fact more in common with black men than with white women, given that their health-related challenges had to do with poverty and racism, rather than gender inequality.

Given these tensions, emancipatory scholars contend that social justice spaces might boil down to nothing more than forums in which only the voices of a vocal few are actually heard. They maintain that top-down, superimposed frames of reference prevent marginalized actors from having a broader influence in agenda setting or implementation.

The differences between rhetoric and actual practice are recognized here, where grand-sounding promises of empowerment may be masking projects that in reality simply enlist people in predetermined ventures where an agenda has already been set (Cornwall, 2003, p. 1327).

## Conclusion

To sum up, in reading earlier studies of emergent groups and empowerment, the process of empowerment seems to happen when people take active part in relief and recovery work, as opposed to passively receiving assistance. However, merging this with ideas from the emancipatory literature, we find that although taking active part is crucial, it is not enough. For empowerment to take place, marginalized communities also need to gain influence over the ends and means of the participatory space. However, as we immerse ourselves deeper into the emancipatory literature, we might expect some trouble along the road, due to the risk that social justice spaces end up exhibiting troublesome differences between rhetoric and actual practice. Social justice spaces where participants span social, economic, and educational differences might – if not well handled – boil down to nothing more than venues in which only privileged people are heard, reproducing the same inequalities that they were initially supposed to change. Whether similar tendencies can be found in post-disaster empowerment processes is however still largely unexplored. Perhaps the urgency and the magnitude of the problems that amount in the wake of a disaster somehow upset or alter these tendencies.

## References

Ahmed, S. (2004). Declarations of whiteness: The non-performativity of anti-racism. *Borderlands*, 3(2). Retrieved from www.borderlands.net.au/vol3no2_2004/ahmed_declarations.htm

Alfrey, L., & Twine, F. W. (2017). Gender fluid geek girls: Negotiating inequality regimes in the tech industry. *Gender & Society*, 31(1), 28–50.

Aptekar, L. (1990). A comparison of the bicoastal disasters of 1989. *Behavior Science Research*, 24(1–4), 73–104.

Bacchi, C. L. (1996). *The Politics of Affirmative Action: 'Women', Equality and Category Politics*. London: Sage Publications.

Baiocchi, G., Bennett, E. A., Cordner, A., Klein, P., & Savell, S. (2014). *The Civic Imagination: Making a Difference in American Political Life*. Boulder: Routledge.

Batliwala, S. (1994). The Meaning of Women's Empowerment: New Concepts from Action. In G. Sen, A. Germain, & L. C. Chen (Eds.), *Population Policies*

*Reconsidered: Health, Empowerment and Rights* (pp. 127–138). Boston, MA: Harvard University Press.

Beady, C. H., Jr., & Bolin, R. C. (1986). The Role of the Black Media in Disaster Reporting. Working Paper No. 56. *Natural Hazards Research and Applications Information Center*. Boulder, CO: University of Colorado.

Black Lives Matter. (2016). Retrieved from www.blacklivesmatter.com

Bondesson, S. (2020). Hurricane Sandy: A Crisis Analysis Case Study. *Oxford Research Encyclopedia of Politics*. Retrieved from https://oxfordre.com/polit ics/view/10.1093/acrefore/9780190228637.001.0001/acrefore-9780190228 637-e-1598

Campbell, C. (2014). Community mobilization in the 21st century: Updating our theory of social change? *Journal of Health Psychology*, 19(1), 46–59.

Carlton, S. (2015). Connecting, belonging: Volunteering, wellbeing and leader-ship among refugee youth. *International Journal of Disaster Risk Reduction*, 13, 342–349.

Chambers, R. W. (1983). *Rural Development: Putting the Last First*. London: Longman Publishing.

Chambers, R. W. (1997). *Whose Reality Counts? Putting the First Last*. London: IT Publications.

Chavis, D. M. (2001). The paradoxes and promise of community coalitions. *American Journal of Community Psychology*, 29(2), 309–320.

Choudry, A., Hanley, J., & Shragge, E. (2012). *Organize: Building from the Local for Global Justice*. Oakland, CA: PM Press.

Cole, T. (2012, March 21). The White-Savior Industrial Complex. Retrieved from www.theatlantic.com/international/archive/2012/03/the-white-savior-industrial-complex/254843/

Cooke, B., & Kothari, U. (Eds.). (2001). *Participation: The New Tyranny?* London: Zed Books.

Cornwall, A. (2002). *Making Spaces, Changing Places: Situating Participation in Development* (IDS Working Paper 170). Brighton, UK: Institute of Development Studies.

Cornwall, A. (2003). Whose voices? Whose choices? Reflections on gender and participatory development. *World Development*, 31(8), 1325–1342.

Cornwall, A. (2016, April). Women's empowerment: What works? *International Development*, 28(3), 342–359.

Cornwall, A., & Goetz, A. M. (2005). Democratizing democracy: Feminist perspectives. *Democratization*, 12(5), 783–800.

Cornwall, A., & Schatten Coelho, V. (2007). *Spaces for Change? The Politics of Citizen Participation in New Democratic Arenas*. New York, NY: Palgrave McMillian.

Correll, S. J. (2001). Gender and the career choice process: The role of biased self-assessments. *American Journal of Sociology*, 106(6), 1691–730.

Crenshaw, K. (1991). Mapping the margins: Intersectionality, identity pol-itics, and violence against women of color. *Stanford Law Review*, 43(6), 1241–1299.

Cudd, A., & Andreasen, R. (Eds.) (2005). *Feminist Theory: A Philosophical Anthology*. Hoboken, NJ: Wiley-Blackwell.

Dacombe, R. (2018). *Rethinking Civic Participation in Democratic Theory and Practice*. London: Palgrave Macmillan.

D'Angelo, C., & O'Connor, L. (2016, December 4). Army Halts Construction of Dakota Access Pipeline. *The Huffington Post*. Retrieved from www.huf fingtonpost.com/entry/obama-dakota-access-pipeline-halt_us_5844882be 4b0c68e04817323

Davis, A. (2016). *Freedom Is a Constant Struggle: Ferguson, Palestine and the Foundations of a Movement*. Chicago: Haymarket Books.

della Porta, D. (2005). Deliberation in movement: Why and how to study deliberative democracy and social movements [Special issue]. *Acta Politica*, 40(3), 336–350.

Dixon, C. (2012). Building "another politics": The contemporary anti-authoritarian current in the US and Canada. *Anarchist Studies*, 20(1), 33–60.

Dovi, S. (2009). In praise of exclusion. *The Journal of Politics*, 71(3), 1172–1186.

Drabek, T. A., & McEntire, D. A. (2003). Emergent phenomena and the sociology of disaster: Lessons, trends and opportunities from the research literature. *Disaster Prevention and Management: An International Journal*, 12(2), 97–112.

Dynes, R. R. (1970). *Organized Behavior in Disaster*. Lexington, MA: Health Lexington Books.

Dynes, R. R. (1987). The concept of role in disasters. In R. R. Dynes, B. de Marchi, & C. Pelanda (Eds.), *Sociology of Disasters: Contribution of Sociology to Disaster Research* (pp. 71–103). Milano, Italy: Franco Angeli.

Dynes, R. R. (1994). *Situational Altruism: Toward an Explanation of Pathologies in Disaster Assistance* (Preliminary Research Paper 201). Newark, DE: University of Delaware, Disaster Research Centre.

Eagly, A., & Carli, L. (2007). *Through the Labyrinth: The Truth About How Women Become Leaders*. Cambridge: Harvard Business Press.

Eliosoph, N. (2011). *Making Volunteers: Civil Life After Welfare's End*. Princeton: Princeton University Press.

Emmanuel, D. (2006). Emergent behavior and groups in postdisaster New Orleans: Notes on practices of organized resistance. In Natural Hazards Center (Ed.), *Learning from Catastrophe: Quick Response Research in the Wake of Hurricane Katrina* (pp. 235–261). Boulder, CO: Natural Hazards Center, University of Colorado.

Ferguson, S. (2016). Dear volunteers in Africa: Please don't come help until you've asked yourself these four questions. *Matador Network*. Retrieved from http://matadornetwork.com/life/dear-volunteers-africa-please-dont-come-help-youve-asked-four-questions/

Fothergill, A., Maestas, E. G. M., & DeRouen Darlington, J. (1999). Race, Ethnicity and Disasters in the United States: A Review of the Literature. *Disasters*, 23(2), 156–173.

Fothergill, A., & Peek, L. (2004). Poverty and Disasters in the United States: A Review of Recent Sociological Findings. *Natural Hazards,* 32, 89–110.

Fraser, N. (1990). Rethinking the public sphere: A contribution to the critique of actually existing democracy. *Social Text*, 25(26), 56–80.

Freire, P. (2005). *Pedagogy of the Oppressed, 30th Anniversary Edition*. New York: Continuum.

Fung, A. (2005). Deliberation before the revolution: Towards an ethics of deliberative democracy in an unjust world. *Political Theory*, 33(2), 397–419.

Fung, A., & Wright, E. O. (2003). *Deepening Democracy: Institutional Innovation in Empowered Participatory Governance*. London: Verso.

Gardner, R. O. (2013). The emergent organization: Improvisation and order in Gulf Coast disaster relief. *Symbolic Interaction*, 36(3), 237–260.

Gould, C. C. (1996). Diversity and democracy: Representing differences. In S. Benhabib (Ed.), *Democracy and Difference: Contesting the Boundaries of the Political* (pp. 171–187). Princeton: Princeton University Press.

Harrald, J. R. (2006). Agility and discipline: Critical success factors for disaster response. *The ANNALS of the American Academy of Political and Social Science*, 604(1), 256–272.

Hayward, C. R. (2004). Doxa and deliberation. *Critical Review of International, Social and Political Philosophy*, 7(1), 1–24.

Helsloot, I., & Ruitenberg, A. (2004). Citizen response to disasters: A survey of literature and some practical implications. *Journal of Contingencies and Crisis Management*, 12(3), 98–111.

Hickey, S., & Mohan, G. (2004). *Participation, from Tyranny to Transformation: Exploring New Approaches to Participation in Development*. London: Zed Books.

Hilmer, J. D. (2010). The state of participatory democratic theory. *New Political Science*, 32(1), 43–63.

Holm, M. (2019). *The rise of online counterpublics? The limits of inclusion in a digital age* (Doctoral dissertation). Uppsala University.

Holm, M., & Ojeda Castro, J. (2018). #Solidaritysiforwhitewomen: Exploring the Opportunities for Mobilizing Digital Counter Claims. *Political Science & Politics*, 51(2), 331–334.

Holvino, E. (2008). Intersections: The simultaneity of race, gender and class in organization studies [Special issue]. *Gender, Work and Organization*, 17(3), 248–277.

hooks, B. (1994). *Outlaw Culture: Resisting Representations*. New York, NY: Routledge.

hooks, B. (2010). *Teaching Critical Thinking: Practical Wisdom*. New York: Routledge.

Houten, D. van & Jacobs, G. (2005). The empowerment of marginals: Strategic paradoxes. *Disability & Society*, 20(6), 641–654.

Jung, C. (2003). Breaking the cycle: Producing trust out of thin air and resentment. *Social Movement Studies*, 2(2), 147–175.

Juris, J. S., Ronayne, M., Shokooh-Valle, F., & Wengronowitz, R. (2012). Negotiating power and difference within the 99%. *Social Movement Studies*, 11(3–4), 434–440.

Karpowitz, C., Raphael, C., & Hammond, A. S. (2009). Deliberative democracy and inequality: Two cheers for enclave deliberation among the disempowered. *Politics & Society*, 37(4), 576–615.

Kruks, S. (2001). *Retrieving Experience: Subjectivity and Recognition in Feminist Politics*. Ithaca: Cornell University Press.

Livingston, R. W., Rosette, A. S., & Washington, E. F. (2012). Can an agentic black woman get ahead? The impact of race and interpersonal dominance on perceptions of female leaders. *Psychological Science*, 23(4), 354–58.

Mahrouse, G. (2014). *Conflicted Commitments: Race, Privilege, and Power in Solidarity Activism*. Montreal: McGill Queen's University Press.

Mansbridge, J. (1976). Conflict in a New England town meeting. *The Massachusetts Review*, 17(4), 631–663.

Mansbridge, J., Bohman, J., Chambers, S., Estlund, D., Føllesdal, A., Fung, A., ... Martí, J. L. (2010). The place of self-interest and the role of power in deliberative democracy. *The Journal of Political Philosophy*, 18(1), 64–100.

McDaniel, J. (2002). Confronting the structure of international development: Political agency and the Chiquitanos of Bolivia. *Human Ecology*, 30(3), 369–396.

Meyer, I. (1995). Minority stress and mental health in gay men. *Journal of Health and Social Behavior*, 36(1), 38–56.

Mosse, D. (2005). *Cultivating Development: An Ethnography of Aid Policy and Practice* (Anthropology, Culture and Society Series). Ann Arbor, MI: Pluto Press.

Neal, D., & Phillips, B. D. (1995). Effective emergency management: Reconsidering the bureaucratic approach. *Disasters*, 19(4), 327–337.

Peacock, W. G., Gladwin, H., & Morrow, B. H. (1997). *Hurricane Andrew: Ethnicity, Gender and the Sociology of Disasters*. New York, NY: Routledge.

Pateman, C. (1970). *Participation and Democratic Theory*. Cambridge: Cambridge University Press.

Phillips, A. (1995). *The Politics of Presence*. Oxford: Oxford University Press.

Phillips, A. (1996). Dealing with difference: A politics of ideas or a politics of presence? In S. Benhabib (Ed.), *Democracy and Difference: Contesting the Boundaries of the Political* (pp. 139–153). Princeton: Princeton University Press.

Phillips, B. D. (1993). Cultural diversity in disasters: Sheltering, housing, and long-term recovery. *International Journal of Mass Emergencies and Disasters*, 11(1), 99–110.

Pilisuk, M., McAllister, J., & Rothman, J. (1996). Coming together for action: The challenge of contemporary grassroots community organizing. *Journal of Social Issues*, 52(1), 15–37.

Randhawa, S. (2016). Poverty porn vs empowerment: The best and worst aid videos of 2016. *The Guardian*. Retrieved from www.theguardian.com/glo bal-development-professionals-network/2016/dec/08/radiator-award-pove rty-porn-vs-empowerment-the-best-and-worst-aid-videos-of-2016

Reinecke, J. (2018). Social movements and prefigurative organizing: Confronting entrenched inequalities in Occupy London. *Organization Studies*, 39(9), 1299–1321.

Ridgeway, C., & Kricheli-Katz, T. (2013). Intersecting cultural beliefs in social relations: Gender, race, and class binds and freedoms. *Gender & Society*, 27(3), 294–318.

Rodríguez, H., Trainor, J., & Quarantelli, E. L. (2006). Rising to the challenges of a catastrophe: The emergent and prosocial behavior following Hurricane Katrina. *Annals of the American Academy Political and Social Science*, 604(1), 82–101.

RUMMET (2013, December 19). Vi vänder oss för första gången till varandra [For the first time, we turn to each other]. *Bang* [online periodical]. Retrieved from www.bang.se/artiklar/vi-vander-oss-for-forsta-gangen-till-varandra

RUMMET (2014, February 25). Att de fortfarande inte fattar [That they still don't get it]. *Bang* [online periodical]. Retrieved from http://rummets.se/blog/chatt-att-de-fortfarande-inte-fattar/

Schneider, S. K. (1992). Governmental response to disasters: The conflict between bureaucratic procedures and emergent norms. *Public Administration Review*, 52(2), 135–145.

Scott, J. (1985). *Weapons of the Weak: Everyday Forms of Peasant Resistance*. New Haven, CT: Yale University Press.

Scott, K., & Liew, T. (2012). Social networking as development tool: A critical reflection. *Urban Studies*, 49(12), 2751–2767.

Seana, S., & Fothergill, A. (2009). 9/11 volunteerism: A pathway to personal healing and community engagement. *Social Science Journal*, 46, 29–46.

Smith, J., & Glidden, B. (2012). Occupy Pittsburgh and the challenges of participatory democracy. *Social Movement Studies: Journal of Social, Cultural and Political Protest*, 11(3–4), 288–294.

Snow, D. A., Soule, S. A., & Kriesi, H. (Eds.) (2004). *The Blackwell Companion to Social Movements*. Hoboken, NJ: John Wiley & Sons.

Solnit, R. (2009). *A Paradise Built in Hell: The Extraordinary Communities that Arise in Disaster*. London, UK: Penguin Books.

Stallings, R. A., & Quarantelli, E. L. (1985). Emergent citizen groups and emergency management. *Public Administration Review*, 45, 93–100.

Stallings, R. A., & Quarantelli, E. L. (2015). Emergent citizen groups and emergency management. In N. C. Roberts (Ed.), *The Age of Direct Citizen Participation* (pp. 71–103). London, UK: Routledge.

Strandh, V., & Eklund, N. (2018). Emergent groups in disaster research: Varieties of scientific observation over time and across studies of nine natural disasters. *Journal of Contingencies and Crisis Management*, 26, 329–337.

Turner, R. H., & Killian, L. (1972). *Collective Behavior* (2nd ed.). Englewood Cliffs, NJ: Prentice- Hall.

Twigg, J., & Mosel, I. (2017). Emergent groups and spontaneous volunteers in urban disaster response. *Environment & Urbanization*, 29(2), 443–458.

Voorhees, W. R. (2008). New Yorkers respond to the World Trade Center attack: An anatomy of an emergent volunteer organization. *Journal of Contingencies and Crisis Management*, 16(1), 3–13.

Whittaker, J., McLennan, B., & Handmer, J. (2015). A review of informal volunteerism in emergencies and disasters: Definition, opportunities and challenges. *International Journal of Disaster Risk Reduction*, 13, 358–368.

Yeich, S. (1996). Grassroot organizing with homeless people: A participatory research approach. *Journal of Social Issues*, 52(1), 112–121.

Young, I. M. (2000). *Inclusion and Democracy*. Oxford: Oxford University Press.

Young, I. M. (2001). Activist challenges to deliberative democracy [Special section]. *Political Theory*, 29(5), 670–690.

# 4    Tales of a Peninsula Shattered and Divided

## A Peninsula Shattered

If weather is the mood of a place, New York City was angry on October 29, 2012. The residents of Rockaway felt this anger bear down on them, as some of them lost their homes, and all of them lost their beloved boardwalk. When you lose essential spaces, when the local coffee shop is closed, or the public library caves in on itself, your sense of community, your attachment to place, or even your sense of yourself can be disrupted (Oliver-Smith and Hoffman, 1999). Rockaway residents' grief over the boardwalk ran deep. One man I talked to felt his sanity at risk for not being able to walk the boardwalk anymore. In 2017, I returned to Rockaway once the boardwalk was restored and rode a bike along the shore. Feeling the warm summer breeze graze my skin and clear my thoughts, I began to understand him.

The hurricane belt – the US Atlantic coast, the Caribbean, and the Gulf of Mexico – is known for its recurring hurricanes. The hurricane season stretches from June to November, with winds sometimes gusting at 300 kilometers per hour. The winds create powerful storm surges that flood low-lying land and barrier islands, often exceeding yearly rainfall levels in only one day. Hurricane Sandy was a rare storm hitting at high tide, making everything even worse. Watching the images of giant waves flooding everything in their path, it is no wonder residents talked about Hurricane Sandy as "the tsunami." Brackish water flooded into basements, utility vaults, and substations. Trees were downed, debris flew around. The storm tore apart the coastlines of New York and New Jersey and long-term electric power cuts followed, leaving seniors trapped in high-rises without functioning elevators for weeks. Forty-three people died. Thousands

DOI: 10.4324/9781003005278-4

became homeless (Chakrabarti, 2013; NASA, 2012; Rosenzweig et al., 2014; Sharp, 2012). In NYC alone the storm was estimated to have cost approximately $19 billion USD (Homeland Security Studies and Analysis Institute, 2013, p. 23). Seven subway lines closed due to flooding and tunnels under the East River were shut down (Rosenzweig & Solecki, 2014, p. 398). Low-income households, public housing tenants, and immigrant communities were among the worst affected (Adams Sheets, 2013; deMause, 2013; Enterprise, 2013, p. 7; Haygood, 2013; Krauskopf et al., 2013, p. 7; Make the Road New York, 2012).

The Rockaway peninsula is a barrier island with a varied ecology of sand dunes susceptible to storm surges (Joseph, 2013). The direct hit of Hurricane Sandy to the peninsula hence caused dreadful consequences. A large number of houses were severely flooded or destroyed altogether. People were displaced to shelters, and schools closed down. Seven Rockaway residents died (Rockaway Waterfront Alliance Report, 2013, p. 11). Gas stations closed, cell phone service fluctuated, and the local newspaper shut down (The Wave, 2013). Survivors were paddling through toxic waste water in the first days after the storm, as sanitation sewage and fecal sludge mixed with debris (Joseph, 2013). The storm demolished parts of the subway track serving the only subway to the peninsula (American Planning Association, 2013, p. 6). Businesses shut down, making it hard for residents to find goods and services while causing local shop owners to lose valuable revenue (American Planning Association, 2013, p. 7). Local companies laid off staff or closed (Gay, 2014; Rockaway Waterfront Alliance Report, 2013, p. 35).

A number of the 402 NYC Housing Authority (NYCHA) buildings, encompassing over 35,000 units, were damaged, many of these located in Rockaway (Furman Center, 2013, p. 4). Basements flooded (rather unfortunate since this is where electrical and heating systems were located), leaving tenants without electricity and heat. One and a half years after the storm, NYCHA residents were still living with mold infestation and unfinished repairs (Colangelo, 2014). The 178,000 units of affordable rental housing of the city faced similar challenges (Furman Center, 2013, p. 5). Some of the problems related to the growth of mold underneath hastily replaced flooring. Damaged homes went then into foreclosure, as displaced owners struggled by the skin of their teeth to pay rent for the new place (Koslov & Merdjanoff, 2013). Although the reconstruction of homes and businesses could have been an employment source to local residents seeking work, many lacked the required licenses to take those jobs.

## Rockaway: From Summer Resort to a Place of Socioeconomic Marginalization

It is only 20 miles between Manhattan and Rockaway, but traveling the distance – even six months after the storm – was a long journey. As the subway train hurled forth from 14th Street, I was surrounded by bearded hipsters, giggling Korean tourists and businesswomen in heels. Once we were on the Brooklyn side, the crowd began to shift. Waiting for the first bus to take us further out to Queens, an elderly black woman sat down on a bench, while I tried to decipher what the two Latino teenage girls were talking about in Spanish. When the last and final bus arrived at Rockaway Beach 59th street, two and a half hours after I left Manhattan, I was the only white person on it.

As with most disasters hitting already vulnerable communities, the devastation of Hurricane Sandy was no bolt from the blue. Anyone who knew a little about Rockaway's long history of social, economic, and political neglect could tell which way the winds were blowing. With Rockaway's beaches and bungalows, its distinctive small town feeling, coupled with the relative closeness to a vibrant metropolitan city center, the peninsula could be a paradise. Yet, it is far from it. There are affluent neighborhoods and beautiful beachfront properties, especially on the Western side. But soaring unemployment, high percentages of people enrolled in social welfare programs, a beach line scattered with poorly maintained high-rise public housing and high levels of criminality, create a situation of socioeconomic marginalization across many communities of Rockaway. Hence, the Hurricane Sandy winds, floods, and rain hit a peninsula already broken, creating a perfect Venn diagram of long-term and acute suffering.

Once a popular summer resort in the early 1990s, crowded with bungalows and hotels, Rockaway was a place in which a small but cohesive population lived; mainly middle- and lower-middle-class neighborhoods of Irish and Jewish communities. By mid-century, the seasonal nature of the area waned (City of New York, 2008). After World War II, the peninsula went through changes – the population grew and ethnic, racial, and class stratification became more pronounced. The isolation of the peninsula made it a suitable space for the city's active relocation of poor communities. Municipal authorities directed marginalized and underprivileged communities to Rockaway, where they lived in public housing projects that were poorly maintained. A large number of group homes for released mental patients and nursing homes for the elderly were built in the 1950s and 1960s (Kaplan & Kaplan, 2003, p. 3). When I spoke to people from more affluent communities in Rockaway, they often used the offensive term "dumping" to describe this process of relocation of NYC's vulnerable communities. Relocation is in line with a general trend in NYC, where focus is on the vibrant city center and outer areas are left to deteriorate. In fact, urban redevelopment programs all over the city have relocated large numbers of poor minority families to peripheral areas of the city (Kaplan & Kaplan, 2003, p. 6).

## A Peninsula Divided

As of 2010, approximately 112,000 residents lived in Rockaway, out of which roughly 40% was African American, followed by 34% white, 21% Hispanic, 2% Asian, and 3% other races. Western neighborhoods are primarily white, mostly third or four generation Irish communities, whereas African American and Latino/Hispanic groups reside mostly on the eastern portion of the peninsula (Rockaway Waterfront Alliance Report, 2013, p. 14). Segregating housing policies and racially skewed urban planning have made parts of Rockaway better off, especially the Irish and Jewish neighborhoods (Kaplan & Kaplan, 2003, p. 5; Kolitz, 2015). This segregation reflects on the unequal political participation and influence between east and west, as well as on the differentiated social vulnerability when Hurricane Sandy hit the peninsula.

After a film screening of a documentary about Rockaway I attended in 2014, a discussion broke out about how the peninsula is divided in East and West. A black woman stood up:

> On the East side, we don't understand politics on a local level. We need civic education; we need to better understand the process, so we can go to the right office with the right complaints. On the West side, they know exactly where to go, and they flood the offices with complaints, and they will not take no for an answer.

Prompted by her comment, I felt the urge to get a broader taste of civic engagement in Rockaway and learn more about Rockaway's more affluent communities. I thus attended a public meeting with a Rockaway local government organ, Queens Community Board 14. The hall room for the meeting was located in one of the local churches, this time a spacious chamber with burgundy broadloom floor. A mixed crowd of roughly 100 people, mostly men, had gathered – old, young, people of color, white people, men in kippahs, well-dressed women, and some men dressed on the verge of trashy. It was not immediately obvious to me that all of them belonged to Rockaway's more affluent communities, based on what I saw. As I took a seat in the back of the room and eavesdropped on the buzzing conversations around me, I sensed that many attendees had a common enemy: the NYC mayor at the time, Bill de Blasio. A dry comment from a public official from the Parks Department, a tired-looking white man in his mid-30s, illustrated the animosity toward the city's administration: "Strange as it may seem, I didn't take this job to be your piñata every night."

The agenda included a discussion about the delayed renovation of the beloved Rockaway boardwalk. The hot topic of the evening, however, became the City's establishment of a shelter for homeless people in Rockaway. People aired some frustrated comments on the issue, and their remarks exemplified the more affluent communities' historical self-understanding of Rockaway as a place where the city administration "dumped" all of the city's social problems. "Out of sight, out of mind," as some of my informants kept telling me when they wanted to explicate the history of Rockaway. State Assemblyman Phillip Goldfeder, at the time representing the 23rd Assembly District in the New York State Assembly (which includes Rockaway), was present at the meeting. His vigorous speech captured the gist of the frustration – "What we need are businesses coming in and they will not while the city keeps dumping on us!" – and received a lot of applause.

After his speech, an unruly discussion broke out among the meeting participants. Wild speculations and rumors about the homeless people inhabiting the shelter were thrown around the room – they were sex offenders, pedophiles, they were panhandling and loitering and "can you believe it, a few of them actually had a barbeque under the bridge the other day!" Such horrible allegation of barbequing notwithstanding, discussions went onto other topics. However, almost two years after the storm, Hurricane Sandy still tangled into almost every topic, from the rebuilding of the boardwalk to pleas for economic support for local organizations.

The concentrations of poverty and unemployment denounced as "dumping" by the more affluent communities in Rockaway have pernicious effects when mixed with spatial and social isolation, resulting in situations where black and Latino children are poorly educated, live in dilapidated circumstances, and have few prospects for employment and thus lower tax revenue bases (Tierney, 2014; Young, 2000). Socioeconomic marginalization in turn triggers criminality, as opportunities shrink and there are less ways to escape poverty in socially and legally legitimate ways (Bauman, 2011). Unemployment is higher on the Eastern side than elsewhere in Rockaway, and among employed residents, most work in low-wage service jobs. More than one-fourth in this area were born in foreign countries, and over half do not have

US citizenship (Rockaway Waterfront Alliance Report, 2013, pp. 15–17). In 2010, the average Rockaway per capita income was $21,172, which is $5,000 less than in Queens and $40,000 less than in Manhattan (American Planning Association, 2013, p. 4). As of 2009, the poverty rate was 20.09%, which is higher than both city- and borough-wide rates (Furman Center, 2015). The 2010 Census outlined a median household income of $38,275, where one-third earned less than $30,000 per year (United States Census Bureau FactFinder, 2015).

Rockaway is home to six public housing developments, with 5,000 units housing approximately 10,000 people, eight group homes, four drug rehabilitation centers, and several Single Room Occupancy hotels (About Section 8, n.d.; Developments of the New York City Housing Authority 2021; Map, 2021; Sullivan & Burke, 2013). The high concentration of high-rise public housing in Rockaway is striking the further east you go.

Access to good education is lacking in the area. As of 2008, approximately every fifth person in Rockaway had no high school diploma and only 22.24% had obtained a Bachelor's degree. Unemployment rates exceed national averages by 7.2%, and its residents receive over twice the borough-wide rate with regard to public assistance programs (Furman Center, 2015). Much of the local economy is seasonal, centering on the summer months when there is a large influx of beachgoers, leaving few opportunities for year-round local employment (Rockaway Waterfront Alliance Report, 2013, p. 35). The lack of local jobs means that many people spend much time traveling in and out of other parts of the city. However, transportation to and from the peninsula is a challenge of its own. The commute to Manhattan is between an hour and a half and two hours in each direction.

Food access is another issue. Locally owned shops are increasingly replaced by franchises or chain stores. The few supermarkets that exist are often expensive, leaving many in the hands of fast food chains for nutritional needs. The number of pharmacies and healthcare facilities are low. Social services in Rockaway are largely absent, and crime and substance abuse rates have steadily increased. School dropouts, infectious diseases, and HIV/AIDS are among the problems that mark this area (Kaplan & Kaplan, 2003, p. 4). Francis, a white woman in her 40s who represented a local community organization, expressed her sorrow over the social and medical situation of the peninsula. She talked about how the city administration built a large concentration of homes in Rockaway, yet never cared to locate any of the necessary medical and social services on the peninsula: "Almost every person who passes me are either sick, dying or mentally ill. There is no help for them out here," Francis sighed.

On a September night, I strolled along the beach with Will, one of my field contacts, a white resident in his mid-40s. The sky was a pink explosion of clouds that draped the horizon. Big oceanfront private homes lined the west side of the shore. Three-story mansions with wide porches, well-tended lawns, and SUVs parked outside, some of them vehicles from the New York Fire Department, American flags waiving in the Atlantic winds. As we walked east, the shore was another thing altogether. Wore-down high-rises, unwelcoming cement blocks that served as group homes for homeless youths, a rickety railroad bridge, uncontrolled weed growing underneath. Affluent and precariat lives, side by side, along the Atlantic. Finally, we ended up chatting with two Latino women enjoying the evening on the porch to their bungalow. Rockaway is just like NYC back in the days, they told me; drug deals, gang shootings, violence. The day before, one of the woman had witnessed a drug deal go down just outside of her house. The other woman told me a story of a Rockaway mother dealing drugs through her kid's diapers. She asked me where I was staying, and when I told her I was on the west side, she said: "People of color who move beyond 116 they'll be out of there in a year. They'll be stared at, even called out. Imagine me even coming up there."

Many of the interviewees talked about the more or less explicit tensions between the affluent, white communities to the west and the poorer communities of color to the east. Chloe, a black woman and resident of Rockaway, asserted that because it is an area with stark economic and racial divisions, there is "a lot of rumoring, scapegoating, gossiping, negative things that happen where people don't want to work together and it can be very hard to build consensus."

Many interviewees however told me that the storm closed some of the gaps between the communities. This mirrors previous studies on how people unite in new ways after disasters, as noted by Oliver-Smith & Hoffman (1999). Likewise, Medwinter (2021) suggests in her study of volunteer-resident relations in Rockaway after Hurricane Sandy that social bonding increased due to the gravity of loss and survival that set in motion affective attachments. Residents I interviewed noted that people in Rockaway came together simply because the need to help each other out was so explicit. They also witnessed an increase in community engagement and saw that people in general became more civically active after the storm. Rachel, a white female resident of Rockaway in her late 20s, elaborated on the turmoil and togetherness that the storm had stirred up and how this mobilized residents of Rockaway to continue the work toward social and political change:

> There was so much upheaval that I knew that big changes are coming and I wanted to be a part of those changes, you know, have

some say in that. And I think a lot of people felt that way. I think a lot of people became much more civically engaged after Sandy. And also just you started to communicate with your neighbors and community members more.

Mark, a white resident in his mid-50s described something similar when he said:

These were always rather insulated communities, people kept to themselves. But the shared trauma of the storm brought people together. And people from all over kept pouring in afterward to help out. People from all races and religions, working beside you doing the shitty job of gutting out your house or what have you. It is pretty hard to stay closed to other people when they help you out like that. So the storm opened people up to each other.

## Conclusion

This chapter has portrayed the historical, social, and demographic dynamics of Rockaway and shown how they were there long before Hurricane Sandy hit the peninsula, causing major damage to housing units, job opportunities, and already strained health services. The Rockaway peninsula, once a holiday summer resort, is a peninsula divided socially and economically. To the more affluent west, Irish and Jewish communities reside, while to the east live mostly socioeconomic marginalized people of color.

In this shattered and divided peninsula, Hurricane Sandy had a pervasive effect on all the factors that created pre-disaster social vulnerability. However, it also brought a sense of community for some of its inhabitants, thus breaking decades of division, if not open suspicion. It was at this moment that the Occupy Sandy movement emerged with the aim not only to provide relief assistance, but to do so in a way that would challenge and eventually disturb entrenched patterns of inequality. The next chapters explore the collaborative relief and recovery work between Occupy Sandy activists and residents of Rockaway, as they embarked on a meandering path of empowerment.

## References

About Section 8 (n.d.). NYC Housing Authority [Webpage]. Retrieved from www1.nyc.gov/site/nycha/section-8/about-section-8.page

Adams Sheets, C. (2013). Staten Island's Hurricane Sandy Recovery Slowly Progressing 6 Months Later. *International Business Times*. Retrieved from www.ibtimes.com/staten-islands-hurricane-sandy-recovery-slowly-progressing-6-months-later-1234807

American Planning Association – New York Metro Chapter. (2013). *Getting Back to Business, Addressing the Needs of Rockaway Businesses Impacted by Superstorm Sandy*. A report produced for the Rockaway Development & Revitalization Corporation.

Bauman, Z. (2011). *Collateral Damage: Social Inequalities in a Global Age*. Cambridge, UK: Polity Press.

Chakrabarti, R. (2013). Hurricane Sandy Impact. *The Huffington Post*. Retrieved from www.huffingtonpost.com/tag/hurricane-sandy-impact

City of New York. (2008, July 23). *City Planning Commission Report*. Retrieved from www.nyc.gov/html/dcp/pdf/cpc/080371.pdf

Colangelo, L. L. (2014, April 22). Arverne East Development Could Get a Jump Start This Year with Housing and Retail Plan. *New York Daily News*. Retrieved from www.nydailynews.com/new-york/queens/project-arverne-east-start-year-article-1.1764108

deMause, N. (2013, January 31). As Sandy Relief Efforts Fade, Crisis Far From Over [Blog]. Retrieved from http://demause.net/category/environment/hurricanes/

Developments of the New York City Housing Authority 2021 Map. (2021). New York City Housing Authority. Retrieved from www1.nyc.gov/assets/nycha/downloads/pdf/nychamap.pdf

Enterprise. (2013, October). *Hurricane Sandy: Housing Needs One Year After* [Research Brief]. Retrieved from https://s3.amazonaws.com/KSPPProd/ERC_Upload/0083708.pdf

Furman Center. (2013). *Sandy's Effects on Housing in New York City*. Retrieved from http://furmancenter.org/files/publications/SandysEffectsOnHousingInNYC.pdf

Furman Center. (2015). Data Search. Retrieved from http://datasearch.furmancenter.org/

Gay, M. (2014, April 15). Madeleine Chocolate Co Seeks Cash to Stay in Rockaway. *The Wall Street Journal*. Retrieved from www.wsj.com/news/articles/SB10001424052702303603904579495962500704646

Haygood, B. (2013, October 30). My Birthday Wish. *The Huffington Post*. Retrieved from www.huffingtonpost.com/ben-haygood/my-birthday-wish_1_b_4176009.html

Homeland Security Studies and Analysis Institute. (2013). *The Resilient Social Network: @OccupySandy, #SuperstormSandy*. Washington, DC: Department of Homeland Security Science and Technology Directorate.

Joseph, A. (2013). Sandy. Retrieved from http://socialtextjournal.org/periscope_topic/sandy/

Kolitz, D. (2015, December 1). The Lingering Effects of NYC's Racist City Planning. *Hopes and Fears*. Retrieved from www.hopesandfears.com/hopes/now/politics/216905-the-lingering-effects-of-nyc-racist-city-planning

Koslov, L., & Merdjanoff, A. (2013, October 29). Materializing Crisis: Housing and Mental Health from Katrina to Sandy. Retrieved from http://socialtext journal.org/periscope_article/materializing-crisis-housing-and-mental-hea lth-from-katrina-to-sandy/#sthash.WcAKiXwW.dpuf

Kaplan, L., & Kaplan, C. (2003). *Between Ocean and City: The Transformation of Rockaway.* New York, NY: Columbia University Press.

Krauskopf, J., Blum, M., Lee, N., Fortin, J., Sesso, A., & Rosenthal, D. (2013). *Far From Home: Nonprofits Assess Sandy Recovery and Disaster Preparedness.* New York, NY: School of Public Affairs at Baruch College, CUNY Center for Nonprofit Strategy and Management, Baruch College Survey Research, and Human Services Council of New York.

Make the Road New York. (2012). Unmet Needs: Superstorm Sandy and Immigrant Communities in the Metro New York Area. Retrieved from www. maketheroad.org/pix_reports/MRNY_Unmet_Needs_Superstorm_Sandy_ and_Immigrant_Communities_121812_fin.pdf

Medwinter, S. D. (2021). Reproducing poverty and inequality in disaster: race, class, social capital, NGOs, and urban space in New York City after Superstorm Sandy. *Environmental Sociology,* 7(1), 1–11.

NASA. (2012). Comparing the Winds of Sandy and Katrina. Retrieved from www.nasa.gov/mission_pages/hurricanes/archives/2012/h2012_Sandy.html

Oliver-Smith, A., & Hoffman, S. M. (1999). *The Angry Earth: Disaster in Anthropological Perspective.* London: Routledge, Taylor & Francis Group.

Rockaway Waterfront Alliance. (2013). Planning for a Resilient Rockaways: A Strategic Planning Framework for Arverne East [Report].

Rosenzweig, C., & Solecki, W. (2014). Hurricane Sandy and adaption pathways in New York: Lessons from a first-responder city. *Global Environmental Change,* 28, 295–408.

Sharp, T. (2012). Superstorm Sandy: Facts About the Frankenstorm. *Live Science.* Retrieved from www.livescience.com/24380-hurricane-sandy-status-data.html

Sullivan, B. J., & Burke, J. (2013). Single-Room Occupancy Housing in New York City: The Origins and Dimensions of a Crisis. *City University of New York Law Review,* 17(1), 113–143.

Tierney, K. (2014). *The Social Roots of Risk: Producing Disasters, Promoting Resilience.* Stanford, CA: Stanford University Press.

The Wave. (2013, October 25). Sandy was here; and the year that followed. Retrieved from www.rockawave.com/pageview/pages/SandyEdition

United States Census Bureau Factfinder. (2015). Retrieved from http://factfin der.census.gov/faces/nav/jsf/pages/index.xhtml

Young, I. M. (2000). *Inclusion and Democracy.* Oxford: Oxford University Press.

# 5  Paths of Empowerment in Disaster Relief

## Cut from the Same Cloth: Occupy Sandy as Part of the Wider Occupy Movement

Occupy Wall Street (OWS) had been sleeping for a year when Hurricane Sandy made landfall on the coasts of New York City. Albeit successfully rallying against the power of the wealthiest "1 Percent," the dismantled network was now licking its wounds, having been worn out both by authorities clamping down on the Zuccotti park encampments, as well as by internal tensions between political fractions of the movement (Gould-Wartofsky, 2015). As news of the storm came tumbling down along with the city's basic infrastructure, activists opened their eyes and listened to the wailing sirens. Tweets started to come through, and dormant Occupy Facebook pages were shook awake by updates on how the storm played right into existing urban inequalities. Occupy was yet again stirring. Activists, permeated in ideals of direct action and mutual aid, started to talk about possible ways to help, reach out to storm-affected people and those willing to lend a hand. One of the first Facebook posts read:

> If you see a need in your community, work to fill it. We will do everything we can to support your efforts! Find like-minded folks, band together, and pool your resources. Start with finding a donation drop off location. Then find a local certified kitchen that will donate their space. Ideal if both are located in the same building. Go door to door. Meet your neighbors. Reach out to local churches, schools, community centers, and businesses. If we can do it, you can too! All Power to the People! Rock on, NYC.
>
> (Occupy Sandy, 2012)

DOI: 10.4324/9781003005278-5

The Occupy movement was a global phenomenon emerging in 2011, with bursts of protests and encampments across the world. Preceded by a wave of protests in the Middle East and the *Indignados* movement in Spain, cities all over the world witnessed a surge in popular protests. In Tunis, Cairo, Madrid, NYC, London, and in cities in South Africa, Australia, Japan, and South America, masses came together to protest. Occupiers around the world were making use of spatial strategies of disruption such as camping in unpermitted places, acknowledging the "spatial dimensions of exclusion and inequality by forcing society to recognize that capitalist accumulation happens in certain places and that these places can be named, located and objected to" (Pickerill & Krinsky, 2012, p. 280). In NYC, OWS took over Zuccotti Park on September 17, 2011, attempting to shine light on growing inequality and grossly disproportionate corporate power (Juris et al., 2012). In a beautifully immersed ethnography of the Occupy movement, Michael Gould-Wartofsky (2015) states the three main points of contention that Occupiers raised: unemployment, housing, and student loan debts. In 2011, 26 million people were under- or unemployed in the United States, among them a disproportionate number of African-American and Latino youth, and 60% of all new hires were low-wage jobs. Alongside the unemployment crisis sat the housing crisis, exacerbated by the economic crash of 2008, with younger, less-educated, and historically disadvantaged minority families bearing the brunt. The third concern was the fact that universities had raised tuition fees to unprecedented levels, a problem stacked on top of the long-standing issue of costly student loans (Gould-Wartofsky, 2015).

The different Occupy sites brought together a mix of protesters: public sector unionized workers, homeless activists, and the most numerous of all: college students and graduates, often members of Far Left political formations. Occupy made use of the slogan "We are the 99%," which came to be a very powerful communication device (Pickerill & Krinsky, 2012, p. 280). The slogan reflected the prevalent frustration that so few seem to hold all the power, while the vast majority lack an equal say in social, economic, financial, political, and ecological processes (Juris et al., 2012). Occupiers' interactions with the surrounding society were sometimes harsh, especially with law enforcement agencies. Accusations of unnecessary repression were many: in Oakland, police were involved in a near-fatal assault, and in other places, camps were aggressively cleared with references to public safety or the need to maintain public order (Pickerill & Krinsky, 2012).

The Occupy movement bore traces of several ideologies such as socialism, Marxism, and anarchism. The North American pocket of

Occupy belonged to an anti-authoritarian current in the contemporary United States left, explicitly positioned against two adjacent forms of organizing: non-profits, who are found to be too integrated into the corporate structure, and centralized party organizations, which are seen as too hierarchical (Dixon, 2012). A few examples of comparable actors, organizations, campaigns, and movements within this current are No One Is Illegal, No Border Network, The Mobilization for Climate Justice, INCITE! Women of Color against Violence, and a range of other community-based racial justice groups, anti-poverty groups, feminist organizations, labor justice groups, environmental justice groups, and radical LGBTQIA groups (Dixon, 2012).

Occupy's Internet-based diffusion was a prominent feature. Most of the action was mediated through a range of online social media and open source software practices. Facebook, Twitter, and blogs functioned as communication engines for the technically and media savvy Occupiers. Combined with more traditional forms of mobilization, this communicational infrastructure lent itself to connecting hundreds of thousands of supporters, share millions of posts, and allowed the movement to engage with large audiences outside the filtering of the mass media (Pickerill & Krinsky, 2012).

However, while Occupiers rallied against external systems of power, the logic of majoritarian populism made it difficult to address internal inequalities. The concept of the 99% was "widely recognized as a powerful semantic coup that frames the Occupy movement as a majoritarian challenge to the disproportionate political and economic influence of an elite few" (Juris et al., 2012, p. 436). But it also made internal differentiation difficult to identify and address, as the concept effectively obscured other types of power relations than class. There was a significant lack of representation of people of color, especially from poor and working-class communities (Juris et al., 2012). The People of Color working group ("POCcupy") was there but found themselves "at once mobilized by the occupation and marginalized by its power dynamics" (Gould-Wartofsky, 2015, p. 98), as they were met with resistance in trying to be heard and funded. There were also accusations of exclusion based on gender, as was shown, for example, in reports of sexual harassment and intimidation that made women feel unsafe and unwelcome in camps (Pickerill & Krinsky, 2012). Although the movement aimed for an equal participation of all, the college-educated, white men among the activists assumed positions of power, influence, and informal roles as coordinators (Gould-Wartofsky, 2015). Structural external forces were at play here as well. Contemporary social justice movements in the United States, especially those that are of an informal, fast-paced character, tend to be composed mainly of privileged people with the economic, social, and cultural resources needed to operate within them. There has been a historical divide between these types of movements – which have attracted mainly white and middle-class activists – and the more formal, communitarian, and often church-based, grassroots organizing that have attracted people of color (Juris et al., 2012).

An integral part of many Occupy tactical choices was the refusal to make explicit demands, although this in itself was an issue of contention between different strands of Occupiers. The anarchists opposed demands that addressed states, parties, or elected officials, as doing so was seen as something that might spur co-optation by political parties or legitimize and recognize the state as an agent capable of implementing these demands (Gould-Wartofsky, 2015). Instead of making demands, activists simply attempted to create alternatives, and so they built tent communities with kitchens, bathrooms, libraries, first-aid posts, information centers, sleeping areas, and educational spaces (Pickerill & Krinsky, 2012). In the parks, occupiers enacted an alternative, "attempting to build, in miniature, the kind of society they wanted to live in" (Schein, 2012, p. 336).

An influential practice within the anti-authoritarian current is the explicit linkages between visionary campaigning and practical solidarity work. In Canada, for example, No One Is Illegal–Vancouver works simultaneously with direct support of migrants facing deportation and campaign work geared toward change of Canadian immigration laws and regulations (Dixon, 2012). In line with this, the Occupy movement carries on in new forms and has morphed into various types of local protest groups such as Occupy Congress (McAuliff, 2012), Occupy Our Homes (Occupy Our Homes, n.d.), or pleas for merging environmental justice with social justice (Athanasiou, 2012).

When Hurricane Sandy bore down on NYC in 2012, Occupiers woke up to the storm, noted its debilitating effects on many poor and marginalized neighborhoods across NYC, and put their skills to use, as I will describe next.

## A Sister Is Born: Occupy Sandy Emerges

Occupy Sandy was cut from the same cloth as the wider Occupy movement. The pattern of the weave was new however. Instead of protesting, activists engaged in direct aid. Experienced as they were in logistics, communication, and organization, the activists were well prepared for the task at hand. The Occupiers had an extensive communication network in place through various social media sites that they used during the Zuccotti Park protests in 2011 and made use of this infrastructure to transmit information in a fast and timely manner. The collective running of an improvised encampment for around two months, in which thousands of people were coming and going on a daily basis, required some heavy organizing work. At Zuccotti Park, activists were fed, sheltered, and basic hygiene needs were taken care of (Gould-Wartofsky, 2015).

Furthermore, Occupy activists knew each other well and as Marcos, a Latino man in his mid-20s and long-time Occupy activist said, the network consisted of people that

> you'd already been in jail with, lot of people that you trusted. And also, we know what we're good at, we know who the computer people are, who can make websites, and blogs and like, we know who are good at analyzing stuff or collecting numbers. And we know the people who are good on the ground.

Referencing the Zuccotti Park protests, another Occupy activist outlined in the video *Occupy Sandy: Mutual Aid Not Charity*: "We were prepared, because we had a year of freaking training, of organizing, mobilizing. Living outside, you know. Of being able to bring it together, from a little, and do a lot" (Elizabeth83486, 2014).

Occupy Sandy activists used social media to mobilize people who wanted to help. After a week, roughly 700 new activist volunteers had been put to work and around 20,000 meals were served daily. At its peak, the network accommodated approximately 60,000 people (Homeland Security Studies and Analysis Institute, 2013). Activists distributed direct aid (food, water, warmth), provided medical care and legal aid, helped with mold remediation and rebuilding, provided psychological help, and were continuously canvassing to assess the needs of the communities they operated in (Homeland Security Studies and Analysis Institute, 2013). They also put up free stores, where storm-affected residents could pick up assets needed (Elizabeth83486, 2014). All of this was sustained with the help of private donations. In only the first six months, Occupy Sandy managed to raise $1,377,433.57 USD (Homeland Security Studies and Analysis Institute, 2013, p. 36).

Occupy Sandy activists identified problematic gaps in the NYC authorities' response to the storm. Critique was directed at first responding agencies for failing to provide residents in outskirt neighborhoods of the city with adequate relief support (Solidarity NYC, 2013; Fox News

Latino, 2012; Killoran, 2012; McCambridge, 2012; Weiser, 2014). The NYC Office of Emergency Management was responsible for coordinating the different city agencies, yet was sidelined by the mayor's office, resulting in a haphazard operation that failed in some aspects, mainly with regard to the evacuation of public housing residents. The so-called Regional Catastrophic Planning Team, which was supposed to bring together emergency managers from other states in the region, also remained inactivated (Liboiron & Wachsmuth, 2013). FEMA received criticism for failing to deliver necessities to affected communities, and a later audit found that its system for distributing supplies like food, water, blankets, and generators was flawed (O'Neil, 2014). The federal government did not approve an emergency measure granting federal money for victims of Sandy until January 2013, almost three months after the hurricane touched land.

Meanwhile, non-profits stepped in to fill the gap in the relief efforts (Krauskopf et al., 2013). Yet, the Red Cross, who is mandated by the federal government to implement relief work in times of disasters (Congressional Charter of The American National Red Cross, 2007) also received criticism for its relief efforts, and for being mostly interested in putting itself forward as a successful agency in the eyes of the public. For instance, vehicles were directed to public press conferences instead of storm-affected areas (Elliott & Eisinger, 2014). The Occupy Sandy activists then responded to what they saw as strikingly unmet needs of NYC's marginalized communities. At the same time, the relief operations were also seen as an opportunity for political mobilization. In a 2013 Al Jazeera article about Occupy Sandy, a high school teacher who helped manage communications during OWS, said: "Many of us believe that the work is inherently political. The storm carves out and lays bare the existing inequities in the city" (Hill, 2013).

## "You Are Not the Protagonist of This Story"

Residents in Rockaway demonstrated frustration and anger toward city authorities and other organizations for failing to provide the peninsula with sufficient relief (Cotner, 2012; Rauh, 2012). In interviews, residents told me about lack of power, lack of generators, lack of pumps for clearing out flooded homes, and lack of information about what was going on, which help was available and what to do about problems, such as mold in the home. Rachel talked about the isolation that ensued as a result of the power outages:

There was no communication, which I think was one of the major failures of the government and the City agencies. Because people didn't know what was going on. We didn't have any phones, we didn't have any computers, we didn't have any connections to media at all, so everything was word of mouth and rumors were spread.

Some of the interviewees spoke about what they perceived as either misguided or outright racist responses from blue light authorities, first responders, and parts of the public. The FDNY went around Rockaway to put out local fires. According to resident Will, however, many of these were street bonfires initiated by residents who needed a place to get warm since they lacked heating due to power outages. The bonfires also functioned as important meeting places for residents who were otherwise isolated without access to electricity (meaning they could not charge their phones or access the Internet). Some interviewees specifically mentioned Beach 116th Street as an important border of this biased response. Beach 116th Street, a buzzing street housing local restaurants, delis, a bank, and some shops, makes up a material as well as symbolic line between Rockaway's Western and Eastern communities. To the west of Beach 116th Street, neighborhoods are predominantly comprised of middle-class homeowners, whereas income levels and life opportunities decrease sharply on the eastern side of 116th Street. Beach 116th Street came up time and again as the line beyond which the bigger relief organizations would not go. Several Occupy Sandy activists suspected that many relief organizations had orders not to go east of 116th street. Occupy activist Leah, a white woman in her late 20s, instead stated that the Red Cross did go east of 116th Street, but brought police escort. New York Police Department (NYPD) personnel used loudspeakers to encourage people to come out on the street and accept help from the Red Cross. But, as Leah dryly noted, in predominantly black areas long plagued by police brutality, heavily critiqued NYPD stop and frisk methods, and with serious mistrust toward the police department, this was a failed strategy.

In contrast, Occupy Sandy gave flesh to the idea of combining direct aid with visionary struggles for justice. The guiding ideal of mutual aid functioned both as a practical tool for relief and as a warning not to re-produce the power dynamics imbued in the relation between helper and helped. Mutual aid is an anarchist political principle, here intended to make storm-affected people active parts in relief operations rather than passive recipients of aid. It represents a reciprocal exchange of resources, based on voluntary services, in which there is no strict line between helper and helped (Turner, 2005). In the documentary film *Occupy Sandy and a People's Relief* (Moyers, 2012), an activist describes mutual aid as the opposite of charity: "Charity has become sort of a dirty word. It suggests hierarchy, I have, you don't, I'll give you my crumbs, but you can't sit at my table." Or as another Occupy Sandy activist puts it, in an orientation for new volunteers:

> You are not the protagonist of this story, they are. You are the supporting cast, helping hand. They know what they need, and you have been helping with that, and they are grateful of that, but it isn't charity.
>
> (Elizabeth83486, 2014)

## Inclusion, Autonomy, and Horizontality

Occupy activists rolled out an extensive relief operation organized through the principles of inclusion, autonomy, and horizontality: everyone who wanted to participate was welcome, anyone who saw a need and had an idea on how to solve the problem had the autonomy to do so, and anyone who felt inclined to take on a leadership role could do so (Bondesson, 2020). In short, and as the next sections will demonstrate, everyone was in, anything was good and everyone was a leader.

### Everyone Is In: Empowerment Through Inclusion

In a movie about public housing in Rockaway called *Home in Housing* (Upp Hunter, 2014), a spokesperson for a community organization gives a speech during a rally outside of the mayor's office: "I didn't have water in my apartment to drink. For weeks! And that was before Sandy!" Later in the film, another black woman talks about how horrible the storm was. Then she pauses for effect and states with emphasis:

> But that's when I started helping out in the community. Right then, because instead of me, oh my god, what I'm gonna do, what I'm gonna do, I said, Oh, I'm at the center, they're giving out water, they need help. Instead of me worrying about myself, I started to help others. That's when I became a member and started to help my community. Sandy did that. So Sandy was good and bad.

Inclusion was an important principle for Occupy Sandy activists. Anyone who wanted to become part of the network was welcome. Especially in response to issues of inequality, Occupy Sandy activists saw it as important to include disaster-affected people in relief work, for them to become active partners in relief work rather than passive recipients of aid. Inclusion as an organizing principle also meant that not only storm-affected people but anyone interested in lending a hand was welcome to do so, without barriers for participation. At first, mostly activists affiliated with OWS showed up. But when established relief organizations had to turn down the large amounts of spontaneous volunteers, as a Red Cross representative stated at a briefing at the Red Cross Head Quarters in NYC, Occupy Sandy became the go-to network for many who were not previously affiliated with Occupy (Analect Films, 2013; Homeland Security Studies and Analysis Institute, 2013).

When I asked activist Leah how she got involved with Occupy Sandy, she told me she had medical training and looked for a place to volunteer after the storm, without any luck. Leah wanted to help, but no organization responded to her contact attempts. When she found out through Facebook that Occupy Sandy was mobilizing, she went to one of the Brooklyn hubs and was immediately sent to an apartment complex with supplies. She expected that someone would meet her there and give her instructions but found out that she was alone. She started to knock on doors and quickly realized that many people in the building were in need of medical attention. Later that day, when she reported back to the Occupy Sandy activists at the Brooklyn site, they said, "It looks like you want to get more involved?" After this day she became increasingly engaged in the organizing, which eventually led her to Rockaway, a place she kept returning to on a daily basis for many months. Leah's story is an illustration of how the principle of inclusion was put to practice, but it also says something about autonomy and horizontality. Leah was immediately included without any barriers to participation, she was then sent out on a mission for which she had to use her own decision-making skills and existing capabilities to solve the problem in an improvised way, and she was encouraged to continue to take the lead.

Occupy Sandy activists invited and built relationships with storm-affected people in marginalized communities. They engaged closely with low-income people, undocumented immigrants, NYCHA residents, and homeless people – populations who often shy away from official organizations (Homeland Security Studies and Analysis Institute, 2013). Inclusion in practice also meant that practically anyone who was willing to collaborate with Occupy Sandy was seen as part of the network.

The story of the community-based organization I volunteered in is a case in point. This was a newly opened community service center with a mission to bring employment opportunities and green technologies to Rockaway. When the storm hit, only two weeks after the offices had opened, the floods completely destroyed the small office spaces. However, when activists teamed up with the director, the community service center re-emerged as a relief hub. Soon the office spaces became a buzzing relief center that provided hot meals, supplies, legal assistance, and allocated incoming volunteers to cleanup efforts throughout the peninsula. Outside the office, activists put up a solar panel truck that provided much needed electricity, so residents lined up to charge their phones. Simultaneously, other activists gutted and started to renovate the offices. Out of this, several other relief hubs came into existence, as activists identified needs and local partners across the peninsula. Everyone who wished to take part in the work was welcome to do so, from Rockaway residents to volunteers from other parts of the city, as well as representatives from official organizations. Activists told me in interviews that several official organizations, such as FEMA and the Department of Health, had individual staffers that collaborated informally with them.

### Everything Is Good: Empowerment Through Autonomy

As the peninsula's hospital and many pharmacies had closed due to the storm, medical needs were dire among Rockaway residents. In a community with many mentally ill persons (e.g., war veterans with post-traumatic stress disorder), not getting hold of medicine can become a big problem fast. A few days into the relief efforts in Rockaway, activist Marcos realized that there was an acute need for a medical aid clinic. He looked across the street from the Occupy Sandy hub, saw an abandoned fur shop and thought, "There, that's where we'll do it!" He put a group of people to work on setting up the space and called a few medically trained anarchists he knew from before. They were looking for places to help and came out the next day. And so, a day after the need was first identified, the medical clinic was up and running, and doctors and nurses filled prescriptions and tended to residents' medical needs, among old furs hanging from the ceiling.

This story speaks to the second principle that characterized the work of Occupy Sandy, namely the ambition to organize the work through an autonomous approach. Occupy Sandy activists were able to go to work doing what they felt compelled to do based on the needs they identified. To organize relief work in an autonomous way was a principle emerging from the Occupy movement's refusal of organizational structure. Deciding what any hub, network, organization, or project should focus on equals power. Rather than having activists enter areas with pre-conceived notions about what the needs and problems are, and how they ought to be solved, autonomy is a deliberate attempt to off-set that power. Autonomy as a principle was also a way to directly oppose the logic of established relief organizations, as illustrated by this statement from an Occupy Sandy activist:

> One of the major distinctions between the State and Occupy, or between FEMA, the Red Cross and other official organizations and Occupy, is that the State seems to have a cookie-cutter approach to what's needed in communities of disaster, and it's not always the case. But they come down from above and drop the pellets and leave. "Here's your stuff. Now, survive." What we've done on the other hand, with the concept of mutual aid in mind, is to go out in those communities, talk to people, knock on people's doors, talk to people in the streets, at stores, at barbecue pits where they serve free food and ask them what they need.
>
> (Elizabeth83486, 2014)

Autonomy thus allowed storm-affected people to shape relief activities according to what they saw fit. Moreover, through putting the principle of autonomy to practice, the activists wanted to make room for improvised solutions to problems. Many of the Occupy Sandy activists were middle-class, educated, yet unemployed people in their 20s and 30s. Since Occupy Sandy was a network in which everyone could go to work doing what they did best, web designers worked on websites, doctors volunteered in spontaneously established medical centers, and lawyers helped with FEMA applications (Homeland Security Studies and Analysis Institute, 2013). Jennifer, an Occupy Sandy activist and white woman in her early 30s who had cooking skills, navigated social media to find out where her capabilities were needed the most:

> My response to everything is food. Like my answer is if there's something wrong I will feed you. So I thought I'd find a kitchen, find someplace where I could make food for people who had lost their homes or whatever. And so Occupy Sandy had cooking opportunities and that's where I signed up or I liked it on Facebook also to find out more about what was going on.

Autonomy as a principle was practically manifested in the swift introduction of new volunteers, as opposed to the more rigid vetting and training recruitment procedures of established organizations. The whole process from registration to orientation to active work took on average 45 minutes, meaning that no time was wasted in putting resources to use (Homeland Security Studies and Analysis Institute, 2013). Autonomy also meant wiggle room for unplanned strategies, such as the emerging color coding system. Activists started to tie yellow ribbons onto their cars or wear yellow armbands in order to identify each other in the sometimes messy gatherings of people, cars, and donations.

The extensive use of social media organically emerged in line with the idea of autonomy. Activists used social media as they saw fit, without any restrictions, to attract and mobilize new volunteers, identify community needs across the city, share information, and fundraise money for relief efforts. They set up Facebook and Twitter accounts, ran a newly established WePay account for donations, and managed the Occupy Sandy webpage (Homeland Security Studies and Analysis Institute, 2013).

The presence on social media sites was in addition to equally intense work on the ground. Occupy Sandy activists continuously scanned the city for ongoing relief initiatives and offered their support to local community leaders. Soon Occupy Sandy hubs popped up in Fort Greene, Park Slope, Williamsburg, Bedford-Stuyvesant, and Rockaway (MacFarquhar, 2012). In the reverberations of the storm throughout the city, networks of first responders sprang up around facilities that had survived functionally from the effects of Sandy. These facilities – for instance, local churches and warehouses – served as hubs for relief and coordination through the efforts of local residents (Williams, 2014). Typically, local religious institutions or shop owners donated space, and Occupy Sandy activists joined in to direct resources to those who needed it (Homeland Security Studies and Analysis Institute, 2013). In Rockaway, as elsewhere in the city, activists teamed up with local first responders in relief hubs from Beach 116th Street to Far Rockaway. Ongoing local efforts were strengthened, their activities and reach amplified, and residents became central figures who took active parts in shaping the work.

Several activities emerged spontaneously in these hubs, as the needs of storm-affected people shifted with time. At first, Occupy Sandy activists provided basic supplies like hot food, water, and cover through blankets and clothes. Specialized capabilities, however, evolved pretty quickly: medical teams were formed to canvass for dead bodies and distribute prescriptions; construction teams were formed that removed water, debris and mold from homes, and renovated them afterward; housing teams emerged that connected survivors with host families; and legal teams started to provide legal advice on insurance issues. The kitchen team continued to provide meals, the communications team managed social media outlets, and the incubation team decided on plans for recovery projects and managed finances (Homeland Security Studies and Analysis Institute, 2013).

One practical instance of the principle of autonomy was the wedding registry. Initially, the network had problems of mismatch between needs and donations. Some hubs received large amounts of donated supplies that were not needed, whereas items that were needed were not donated. With the autonomy they had to immediately go to work, some Occupy Sandy activists set up an online space for donations using wedding registries on Amazon.com. Here, storm-affected people could list items that reflected their actual needs, which guaranteed that the right items were donated and delivered to the right locations. People who wanted to donate were then able to purchase supplies like batteries, dehumidifiers, space heaters, generators, and hygiene products (Homeland Security Studies and Analysis Institute, 2013).

The story of Dev, an Indian-American middle-aged man, is an illustration both of autonomy and of remarkable moral commitment. A few days after the storm had hit, Dev took his bike (since roads were closed) and embarked on the one and a half hours bike ride out to Rockaway. When talking to people on the ground he realized that they needed food. He initiated a small hot meal operation with the help of his extended family. For days on end, Dev's family cooked, and he delivered by bike. In the spirit of autonomy, and as needs shifted, Dev went from providing hot meals to working with home repairs. He ended up quitting his job to devote his time to helping people renovate their homes. For a year, Dev did daily trips to Rockaway. He became a social point of contact for many isolated, elderly people who had lost their homes. Dev also provided them with much needed psychosocial support for which he had no prior training. In interviewing Dev, tears welled up both his eyes and mine, when he told me of elderly, vulnerable people, isolated and left to fend for themselves after a disaster that had struck them to the bone:

> They've never seen this kind of devastating disaster in their lives. What they can do is just hug you and cry. It is not easy. I've seen people hang themselves after Sandy. I've seen families frozen to death, in the early time of Sandy, because they had no heat. Husband died, frozen to death, they were married for 46 years. Wife couldn't take it, she cried and she cried, and then she died too. [...] And they talk to me and they tell me they don't want to live anymore. They tell me, "If I'm not here tomorrow, its ok, it's

not because you couldn't help me, I am thankful for your time and help." And so I've got to run to them and talk to them, and sit with them for hours and hours.

### *Everyone Is a Leader: Empowerment Through Horizontality*

Within the anti-authoritarian social justice movement – particularly in anarchist currents, the principle of horizontality is meant to counteract social inequalities between privileged and oppressed people. Social justice forums tend to explicitly oppose hierarchical organizational infrastructures, where decision-making is concentrated in the hands of a few people, often privileged in terms of race, education, and class. In line with these principles, Occupy Sandy activists were skeptical toward charity-based relief approaches, which they saw as diminishing for the people who receive the aid because it reproduces power imbalances between helper and helped. Instead, Occupy Sandy activists aimed for a horizontally based network, in which storm-affected people could participate on equal terms. This was seen as the opposite of what they perceived as ineffective – not to mention morally questionable – top-down structures of traditional organizations. Horizontal relations, most prominently manifested in shared leadership, are believed to breach such inequality by making place for leaders among the oppressed. Shared leadership is believed to prevent domination by a few, since it prevents a situation in which any one person is granted too much control. Troy, a black man in his 30s, resident of Rockaway and an emerging leader within Occupy Sandy, told me in an interview that people can become

> power-hungry and they may begin to make decisions that are not conducive to the group. They become too empowered. We've seen it happen ... So we try to keep the power balanced, we try not to let anyone person be too powerful in the group.

Another important rationale for horizontality was the goal to empower residents to take on leadership roles, especially since the Occupy Sandy activists were well aware that they themselves were privileged in relation to the residents. Through the principle of horizontality, power imbalances that may find their way into the network were challenged. Role allocation in the Occupy Sandy network developed organically: some people took on leadership roles and some were content with implementing other people's decisions. In this way, there were many local leaders who simultaneously took on various responsibilities. Leaders would communicate in ways that differed greatly from hierarchical organizations' vertical patterns of communication. According to a few interviewees, horizontality activated otherwise marginalized people to take active part in relief work, instead of being at best passive recipients of aid, and at worst completely on their own. One Occupy Sandy activist detailed what was happening in Rockaway:

> It's a crazy disaster zone; it looks like New Orleans looked after Hurricane Katrina, but people there are amazing. And being part of that is incredible. The people from that neighborhood, most of them, that have stepped up and are now organizers, and directing traffic, and running security, and going door-to-door canvassing, looking for elderly that need help, those people never did something like that before.
>
> (Dwayne, 2013)

As elsewhere in the city, residents in Rockaway acted as important first responders in their neighborhoods immediately after the storm. They stepped up to check on their neighbors, donate supplies, and aid in getting resources out to people in need. Monique, a black woman and Rockaway resident in her early 30s, initiated a small distribution center from the church she had previously evacuated. After a first few days of isolation, she contacted some Occupy Sandy activists who wanted to support her efforts. Although a humble person, I could still hear the pride in her voice when she talked about how she became a central figure in the relief work after that point. Occupy activists sent both resources and volunteers her way, amplifying both the ongoing relief work she was already undertaking and her role as an emerging local leader. Monique's story is an example of the way Occupy activists teamed up with local first responder initiatives and offered support to local community leaders. Ongoing local efforts were fortified, their activities and reach amplified, and new hubs emerged with residents taking on central leadership roles.

Occupy Sandy activist Leah noted that the storm brought about a strengthening of hyper-marginalized community residents. Although many people were hurt and suffered from damaged houses, lack of heat, food, and medicines, even the most vulnerable individuals made efforts to help out, she said. From her viewpoint, it seemed like some individuals with mental illness or substance abuse problems were able to step up and take on responsibilities. She talked about one man who suffered from alcoholism but sobered up after the storm and helped in directing traffic. Leah, amazed by this, attributed it to the whole feeling of togetherness and community that the storm and the organizing efforts had created.

Occupy Sandy's relief work provides an illustration of what untrained, ordinary people – some of them directly affected by the disaster – can accomplish in networks organized around the principles of inclusion, autonomy, and horizontality. Adam Greenfield, a NYC-based writer with previous experience from the US Army, himself part of the Occupy Sandy relief work, wrote that he had "rarely seen such highly functional order assemble itself so rapidly" (Greenfield, 2013). People from marginalized communities, heavily affected by the storm, were empowered in the process. Not only did they step out from the role of helped and into the role of helper, they also actively participated in autonomously shaping the work according to the local needs they identified. In this way, ongoing local relief activities were reinforced and local leaders were encouraged and supported.

## Conclusion

This chapter explains how the Occupy Sandy movement had its origin in the wider Occupy movement that took to the streets of NYC and other world cities demanding deep structural change. The Occupy movement, with wide logistic, communication and mobilization experience, took relief matters in their own hands when they saw that public agencies and non-profits failed to deliver to people in need, mostly those living in already marginalized areas of the city. Cut from the same cloth as OWS, Occupy Sandy based its approach to direct relief aid on three basic principles: inclusion, autonomy, and horizontality. As Occupiers had tried to create the world they wanted to live in through their encampment in Zuccotti Park, Occupy Sandy activists created a network of relief that challenged the often hierarchical relation between helper and helped. Hence, Occupy Sandy's means and aims were directed at empowering residents in marginalized areas by letting them assess their own needs and supporting them to organize and implement relief work as leaders rather than as passive recipients.

However, as the wet and cold of the hurricane season slowly turned into warmer winds of spring, the immediateness of the relief work transitioned into long-term recovery. When I stepped off the bus for the first time in Rockaway in May 2013, and attended my first meeting with Occupy activists and Rockaway residents, I got to observe how this transition sparked conflicts and tension, as described in the introductory chapter. If the relief phase had shook things loose, had presented an opportunity to break with problems of inequality, something else seemed to be going on in the long-term recovery phase. What were the strains that made things go south? What happened with the rosy promise of empowerment that had been the story of the relief work? In Chapter 6, I delve into these questions.

## References

Analect Films. (2013). *We Got This (Occupy Sandy)* [video]. Vimeo. Retrieved from https://vimeo.com/53155081

Athanasiou, T. (2012). Linked Fates: Occupy and the Climate Negotiations. *Eco Equity*. Retrieved from www.ecoequity.org/2011/11/high-speed-history

Bondesson, S. (2020). Hurricane Sandy: A Crisis Analysis Case Study. *Oxford Research Encyclopedia of Politics*. Retrieved from https://oxfordre.com/politics/view/10.1093/acrefore/9780190228637.001.0001/acrefore-9780190228637-e-1598

Congressional Charter of The American National Red Cross. (2007). Retrieved from www.redcross.org/images/MEDIA_CustomProductCatalog/m4240124_charter.pdf

Cotner, M. (2012, November 5). Feeling forgotten in the Rockaways after Hurricane Sandy. *Brownstoner* [online digital publication]. Retrieved from http://queens.brownstoner.com/2012/11/feeling-forgotten-in-the-rockaways-after-hurricane-sandy/

Dixon, C. (2012). Building "another politics": The contemporary anti-authoritarian current in the US and Canada. *Anarchist Studies*, 20(1), 33–60.

Dwayne, H. (2013). *Occupy Sandy Relief* [Video]. Vimeo. Retrieved from https://vimeo.com/53203990

Elizabeth83486. (2014, May 1). *Occupy Sandy: Mutual Aid Not Charity* [Video]. YouTube. Retrieved from www.youtube.com/watch?v=NMxuE2T6nFg

Elliott, J., & Eisinger, J. (2014, December 11). How Fear of Occupy Wall Street Undermined the Red Cross' Sandy Relief Effort. *ProPublica*. Retrieved from www.propublica.org/article/how-fear-of-occupy-wall-street-undermined-the-red-cross-sandy-relief-effort

Fox News Latino. (2012). Fed Up Staten Island Residents Organize Own Relief Efforts, Want Election Recount. Retrieved from http://latino.foxnews.com/latino/news/2012/11/08/fema-fed-up-staten-island-residents-organize-own-relief-efforts-want-election

Gould-Wartofsky, M. A. (2015). *The Occupiers: The Making of the 99 Percent Movement*. New York: Oxford University Press.

Greenfield, A. (2013, February 6). A Diagram of Occupy Sandy. *Urban Omnibus*. Retrieved from https://urbanomnibus.net/2013/02/a-diagram-of-occupy-sandy/

72    *Paths of Empowerment in Disaster Relief*

Hill, E. (2013, October 27). One year after Sandy, the flood of Occupy volunteers recedes. *Al Jazeera America*. Retrieved from http://america.aljazeera.com/articles/2013/10/27/one-year-after-sandythefloodofoccupyvolunteersrecedes.html

Homeland Security Studies and Analysis Institute. (2013). *The Resilient Social Network: @OccupySandy, #SuperstormSandy*. Washington, DC: Department of Homeland Security Science and Technology Directorate.

Juris, J. S., Ronayne, M., Shokooh-Valle, F., & Wengronowitz, R. (2012). Negotiating Power and Difference within the 99%. *Social Movement Studies*, 11(3–4), 434–440.

Killoran, E. (2012). Occupy Sandy: Occupy Wall Street Finds New Purpose in NYC Hurricane; Old Tensions Remain. *International Business Times*. Retrieved from www.ibtimes.com/occupy-sandy-occupy-wall-street-finds-new-purpose-nyc-hurricane-old-tensions-remain-887124

Krauskopf, J., Blum, M., Lee, N., Fortin, J., Sesso, A., & Rosenthal, D. (2013). *Far From Home: Nonprofits Assess Sandy Recovery and Disaster Preparedness*. New York, NY: School of Public Affairs at Baruch College, CUNY Center for Nonprofit Strategy and Management, Baruch College Survey Research, and Human Services Council of New York.

Liboiron, M., & Wachsmuth, D. (2013, October 29). The Fantasy of Disaster Response: Governance and Social Action during Hurricane Sandy. *Social Text Online*. Retrieved from http://socialtextjournal.org/periscope_article/the-fantasy-of-disaster-response-governance-and-social-action-during-hurricane-sandy/

MacFarquhar, L. (2012, December 3). Occupy Sandy. *The New Yorker*. Retrieved from www.newyorker.com/news/news-desk/occupy-sandy

McAuliff, M. (2012, January 17). Occupy Congress' Protesters Swarm Capitol Hill To Represent The 99 Percent. *Huffington Post*. Retrieved from www.huffingtonpost.com/2012/01/17/occupy-congress-capitol-hill_n_1211294.html

McCambridge, R. (2012, November 7). Why Was Occupy Wall Street Being Occupy Sandy? *Nonprofit Quarterly*. Retrieved from https://nonprofitquarterly.org/policysocial-context/21329-why-was-occupy-wall-street-being-occupy-sandy.html

Moyers, B. (2012, November 16). *Hurricane Sandy and a People's Relief* [Film]. Billmoyers.com. Retrieved from https://billmoyers.com/content/peoples-relief/

Occupy Our Homes. (n.d.). Retrieved from http://occupyourhomes.org

Occupy Sandy. (2012, May 30). Occupy Sandy's first Facebook post [Facebook Status Update]. Retrieved from www.facebook.com/OccupySandyReliefNyc/

O'Neil, E. (2014, October 1). FEMA's system for delivering supplies after disasters like Sandy is flawed, audit finds. *NJ.com*. Retrieved from www.nj.com/news/index.ssf/2014/10/audit_finds_flaws_with_fema_system_for_delivering_supplies_after_disasters_like_sandy.html

Pickerill, J., & Krinsky, J. (2012). Why Does Occupy Matter? *Social Movement Studies*, 11(3–4), 279–287.

Rauh, G. (2012). Mayor Greeted By Upset Residents in Rockaway Beach. *Time Warner Cable News*. Retrieved from www.ny1.com/content/politics/politic al_news/171774/mayor-greeted-by-upset-residents-in-rockaway-beach

Schein, R. (2012). Whose Occupation? Homelessness and the Politics of Park Encampments. *Social Movement Studies*, 11(3–4), 335–341.

Solidarity NYC. (2013). *Growing a Resilient City: Possibilities for Collaboration in New York City's Solidarity Economy.* Retrieved from http://solidarity nyc.org/wp-content/uploads/2013/02/Growing-A-Resilient-City-Solidarity NYC-Report.pdf

Turner, F. J. (2005). *Encyclopedia of Canadian Social Work.* Waterloo, Ontario: Wilfried Laurier University Press.

Upp Hunter. (2014). Living Rockaways: 3 Stories. Hunter Colleague: Urban Policy and Planning. Retrieved from www.hunterurban.org/studio/living-rockaways-three-stories

Weiser, B. (2014, September 30). New York City to Alter Preparations for Disabled in Disasters. *New York Times*. Retrieved from www.nytimes.com/2014/10/01/nyregion/city-to-alter-preparations-for-disabled-in-disast ers-.html

Williams, E. (2014). *Social Resiliency and Superstorm Sandy: Lessons from Community Organizations.* Report from the Association for Neighborhood and Housing Development.

# 6   Paths of Post-disaster (Dis-)Empowerment

## Relief Turns to Long-Term Recovery

As the immediate relief needs after Hurricane Sandy in NYC started to decrease, alliances of community groups, labor unions, faith-based organizations, and environmentalists came together to demand a just and sustainable rebuilding so that the "tens of billions of dollars do not end up in the hands of the same people that created those injustices" (Liboiron, 2013). Attention was now increasingly given to people in poorer areas, low-income households and tenants in public housing, as well as immigrant communities, all of whom had a difficult time returning to a decent life (Adams Sheets, 2013; deMause, 2013; Enterprise, 2013; Haygood, 2013; Krauskopf et al., 2013; Make the Road New York, 2012). Spokespersons from a range of alliances and interest groups asserted that the storm exposed deep inequalities based on income, race, housing, and immigration status (ALIGN, 2013; Jaffe, 2013; Murphy, 2011; Rebuild by Design, 2013; Rohde, 2012; Solidarity NYC, 2013).

Against this backdrop, the Occupy Sandy network extended its work beyond the relief phase to deal with long-term issues of social justice in relation to the storm. Occupy Sandy activists identified a number of long-term recovery projects, using the momentum from the relief work to do so. Across the city, activists set up various hubs where they advocated a recovery process that would address issues of social justice and mobilize and empower storm survivors from marginalized groups. One of these hubs was a grassroots community group located in Rockaway. Occupy Sandy activists approached residents with whom they had worked during the relief phase and invited them into this group. The first step was to sensitize the residents to issues of power imbalances and train them to become leaders and organizers. Through dialogue exercises, the group formed ideas

DOI: 10.4324/9781003005278-6

around the kind of change the residents wanted to see in their community after the storm. A few smaller working groups with more concrete focus gradually grew out of these initial discussions. One working group focused on building environmentally sustainable community projects, one planned protests against the NYPD stop and frisk methods, one focused on creating a worker's cooperative, and one group focused on land-use and disaster-triggered gentrification.

The land-use group eventually came to be the main priority of the Rockaway hub. They engaged in a process of urban development that took place in Rockaway after Hurricane Sandy. The storm had triggered an official revisit to earlier plans for a piece of land called Arverne East (Ellefson, 2014). Occupy Sandy wanted to put pressure on developers to take the needs of Rockaway's poorest into account when planning for this new housing area. For this purpose, they started to build a coalition of grassroots organizations in Rockaway with the aim to create a Community Benefits Agreement (CBA). A CBA is a private agreement between developers and community coalitions. If successfully negotiated, a CBA can enforce legal protection for low-income individuals by taking land off a speculative market, and control land-use decisions through community ownership. In such processes, low-income communities benefit from development (Baxamusa, 2008; Salkin & Lavine, 2008). The main tenets of the draft CBA were deeply affordable housing, demands for minimum wages and that jobs should go to local residents primarily, and green technology/resilience (CBA, 2015). In addition to the work on the CBA, the group also mobilized around issues of climate change and environmental justice in preparation for the upcoming People's Climate March. The march took place in September 2014 and gathered roughly 300,000 demonstrators in NYC. Activists and residents in the hub organized different types of events, such as fundraising parties and information and networking events.

The emancipation that took place within this Occupy Sandy long-term recovery hub was a multifaceted and complex process that moved along a meandering path of both empowerment and disempowerment. A few of the residents that I interviewed talked about how they were empowered by their participation. They increased their knowledge on urban development, climate change, and environmental justice and realized that these were issues of importance to their lives. They also expressed how they became more experienced in organizing meetings, facilitating exercises, and mobilizing for demonstrations and events. Other residents were less excited, some to the point of real frustration. They pointed to how the activists exhibited belittling attitudes toward the residents and talked about how the activists failed to acknowledge the situated marginalization of being affected by the storm. Some perceived the activists as speaking with forked tongues. The residents evinced non-transparent agendas hidden behind a rhetoric of local ownership. In the remaining chapter, the story of post-disaster (dis-)empowerment in this Rockaway Occupy Sandy hub will be discussed.

## Empowerment Through Participating, Organizing, and Learning

The calendar marked November 3, 2013, and winter was coming. Occupy activist Aisha, a woman of color in her late 20s, and I were making our way out to Rockaway from Brooklyn in her dad's SUV. Sitting in the passenger seat, I listened to her talking on the phone with a few other activists, planning the meeting we were driving to. The meeting was held in a semi-abandoned warehouse, and when we arrived, I helped Aisha to set up tables with food and organize sign-up boards. Another Occupy activist put up a table in the back with crayons and papers, and two children gathered around the table while their mother Rachel helped herself to some food. In a later interview, Rachel explained why she was drawn to Occupy Sandy. She said that Sandy shone a light on existing problems of the area, and that all of these problems needed to be addressed for "real recovery" to happen. At Rachel's first meeting, she was attracted to the format of the meeting and how different it was from anything she had experienced before. She thought that people were listening carefully to each other, that the exercises were creative in style, and that the discussions were an opportunity to delve deeper into the kind of changes that the community wanted to see in the wake of the storm. Another important feature was the childcare provided, since she was a mother of two.

Over some stewed kale and potato, I chatted with Sylvette, a middle-aged black woman in sharp-looking attire and golden earrings, who had come to an Occupy Sandy meeting for the first time. She told me a story of how she was the only black female student at Rutgers University back in the days, and how she struggled with Klu Klux Klan influences among faculty members. After a while, resident Monique initiated the meeting and invited everyone to take a seat in the half circle of chairs. Monique told me in an interview that before the storm, she paid no attention to political activism. However, as she worked tirelessly to provide relief for her family and her community, she developed an interest, partly due to the relationship she built with Occupy Sandy activists. When asked by one of them if she wanted to join the hub for more long-term work, she became curious. The friendly low-key invitation was what drew her in, she said, and soon she became part of the core coordinators team, facilitating exercises and introducing newcomers to the meeting rules and principles. When everyone was seated, Monique explained some of the ground rules:

> We want you to be mindful that we have a diverse group of people present, we come from different backgrounds and might not always agree with each other. We ask you to be mindful so as not to be offensive to anybody based on what you know or do not know about the other participants.

Occupy activists from the OWS network had developed a social signal system that Monique then explained. Whenever a participant heard a statement that they agreed with, they would discretely tap their fingers together to show support, producing a low whisper with their hands. If a participant felt that a statement was offensive or misdirected, they would say "ouch" to signal this. Monique then asked everyone to introduce themselves by name, and a brief statement about why they had come to the meeting. There were 16 people present: nine of them were Occupy activists and Rockaway residents already involved with Occupy, and seven were Rockaway residents attending their first or second meeting. Some of the residents talked about how they were there because the voices of their community were not often heard. They talked about how Rockaway was a forgotten place. "It is a place left behind, but we are going to start catching up now," as Sylvette said. She also referred to God, as did some of the other residents. I also introduced myself and told them why I was there and that I would be asking people for interviews. Monique filled in and encouraged people to take the opportunity to talk to me about empowerment.

Occupy Sandy meetings were often creative in style and had an overall fluidity about them. People came and went, exercises shifted into discussions, no exact timeframes were kept, and often enough they would not end with any tangible results. This day was no exception. Aisha took the floor and began by asking us to close our eyes and breathe in and out a few times: "But don't expect this to be one of those bad ass mediations that you've done in yoga, this is way less sophisticated." She prompted us to sit in silence and think about a conversation we had with someone that brought that person closer to something important.

We were then split up in pairs to discuss the situations we had been thinking about. I was paired up with resident Troy. Afterward Aisha gathered everyone in the half circle again and started to take comments from the groups about some of the lessons learned from the situations we had discussed. Occupy activist Eliana said that it was important for her to see that she was helping because it made her feel stronger. Resident Will blurted out: "It's mutual empowerment!" to which the other participants tapped their fingers. Aisha then said: "But when I'm trying to get my brother to do something he does not want to do, that's not mutual empowerment!" and got some hearty giggles. When Troy was asked what had come up for him during the exercise, he chuckled: "I was talking to this person and she was a little weird and was taking notes all the time and I was wondering what was up with that."

Alicia, a black Rockaway resident in her late 40s, who also attended the meeting and whom I later interviewed, talked about the friendliness of the activists, and how welcoming they were to everyone who wanted to join:

> Even though I've only been a few times, I feel like I'm making friends. And before, I'd just go to my apartment. But since Sandy I've made new friends. So you know, I'd see Hannah and she goes, "Ooh, let me give you a hug". You know. So the level of acceptance and the welcoming is really nice. And I feel more confident in speaking up. Even day one I felt confident. I don't know, I can't tell you why but from day one I felt confident in speaking up and asking questions and finding out more and more.

As with Monique, the activists invited Rockaway residents who had been collaborating with them in the relief work and had shown some kind of initiative or leadership. In the first months, the open meetings of the Occupy Sandy hub saw between 40 and 60 people show up. Activists and residents worked across a range of social positions, with varying social, racial, economic, and educational backgrounds. Among the residents, most were low-income persons of color, already marginalized socially and economically, struggling to get back on their feet after a storm that had ripped their neighborhood to pieces. They had little experience of the type of social justice work in which they were invited to take part. The Occupy Sandy activists were mostly young people, mostly non-affected by the storm, were mostly white, mostly educated, and they had organizational skills as well as economic funds that they controlled. They were also mostly familiar with each other from having been involved in Occupy Wall Street.

Both residents and activists thought of the diversity of the group as a strength. It was seen as internally beneficial because the various social positions that the members inhabited were complementary to each other. Residents from the eastern, less affluent end of the peninsula were seen as knowledgeable based on their understanding of "the hood," from the activists' perspectives. To know how people think in these communities of Rockaway was seen as a valuable tool. Some of the residents pointed to how the organizing experiences of the activists were helpful for the group. Most of the interviewees also thought of this diversity as a positive factor for the external work. It enabled the group to reach out to larger parts of the peninsula, as they were a group that in its own configuration spanned some of the important social divides of the area.

Apart from meetings, the group also organized public events, actions, and demonstrations. One of the first actions of the group was a strategic disruption of a city supported project called NYC Special Initiative for Rebuilding and Resiliency (SIRR) (SIRR, 2012). SIRR was a city-initiated formal process that presented a coordinated series of workshops in which 320 community-based organizations and businesses took part, with the stated aim to engage citizens in the transformative plans of rebuilding NYC in a resilient way. Occupy Sandy activists perceived SIRR to be a cynical attempt at appearing to be listening to community voices and thereby gaining traction for urban planning solutions that were in the making pre-Sandy. They decided to go to the Rockaway meeting, spread out in the working groups and then disrupt the discussions by calling attention to the misguided way in which the city was handling the recovery after the storm. Apart from successfully disrupting the meeting, other residents in attendance became aware that Occupy Sandy had set up a long-term camp in Rockaway.

Later, in August 2014, the group organized a so-called Climate Justice Teach-in and Bash event on the eastern side of Rockaway, aiming to mobilize people for the upcoming climate march. Participatory art projects were in full swing when I arrived, and dance and music performances were on the agenda for the day. Themes of discussion were local farming, renewable energy, and grassroot campaigns. A bouncy castle was wiggling under the weight of playing children. Three activists were giving a presentation. They were protesting the planned building of a gas pipeline outside of Rockaway's shore and had painted colorful images on large cloth canvasses portraying greedy white men in suits receiving money from the fossil fuel industry. The event was in preparation for the NYC Climate March of 2014, where eventually 400,000 people flooded the sunny September streets of Manhattan. I was there as well, marching with the Rockaway crowds. All around me people were dancing, shouting, chanting, and cheering. Black boys and men from Rockaway were holding up cardboard signs painted as great waves and lifebuoys with the text "Shorefront Communities Unite." Every once in a while, someone chanted, answered by the audience: "What do we want? Climate Justice! When do we want it? Now! If we don't get it? Shut it down! If we don't get it? Shut it down!"

### *"There Were Pieces of the Puzzle that Weren't Written in the Books"*

Despite the more showy actions, rallies, and demonstrations, most of the initial work of the group was focused on a sort of Freirean consciousness-raising. The meetings consisted of different exercises and group discussions around organizing, most often planned by the activists. Occupy Sandy activist Matthew, a white male in his mid-20s, described it as a period of raising awareness after the immediate effects of the storm had rung out. Participants started to critically asses what had happened in their community in the wake of the storm, why responsible government agencies did not fulfill their responsibility, and how all of this was related to class and race stratification. Most of the residents talked about their experiences from this training period as positive. Going through it, they were able to get political training around issues that they perhaps had thought about but never had the chance to reflect deeper on. Resident Troy talked about seeing new pieces of the puzzle:

> I was telling myself that if I read a lot of books, if I educate myself to what was going on around me, I'll be able to empower myself and then I would take the lead on the things that I wanted to do. But as I went on I realized that there was more to it. I started seeing that there were pieces of the puzzle that weren't written in the books. But when I met the people I was able to identify the missing parts that I didn't have. It provided direction of what I wanted to do and where I wanted to go. They provided small tools for me to actually do that.

As Troy, other residents welcomed the Occupy Sandy activists, experienced their work as an act of solidarity and used their participation as a vehicle for collective and individual capacity-building. The residents who experienced empowerment were the same ones who were encouraged by the activists to become part of the core coordination group. As part of this group, they took active part in shaping the agenda of the organization, participated in the implementation of set strategies and had a say in how the organization ought to structure its work. They stated in interviews that they learned new organizing skills, such as how to facilitate workshops, or plan political actions and they got a better understanding of issues they deemed important, such as how urban planning processes are structured and how political pressure can be put on developers and politicians.

The fluidity of the meetings was a source of positive remarks from some residents. Resident Alicia experienced the meetings format as a source of energy:

> I like the circle because a lot of times when we go to meetings there are tables. But I like the circle because it is more conducive to contribution, and ideas and generating conversations.

The meetings sometimes lacked tangible results. Not every meeting had a clear purpose, and some of the residents talked about how they would leave meetings without a sense of any clear results. But this was not always seen as a bad thing; it was also interpreted as a form of community building that had to take time. Troy noted that the constant role playing, exercising, and open-ended discussions around political issues strengthened the group and resulted in increased feelings of trust and mutual recognition among participants.

One night in August 2014, in a rundown industrial building in Brooklyn, I tagged along to a fundraising party for the climate march. We made our way up the dingy stairs and hallways, following the heavy beats emanating from a hall room somewhere. The party was in full swing when we found the room, with people dancing and milling about. Along the walls, tables had been set up with fliers and sign-up boards for jotting down email addresses. Later, a white Occupy activist I had never seen before took the stage and made an argument about the connections between racial injustice and climate change vulnerability. She added that Rockaway residents are integral to the planning of the march since it is "people from shorefront communities that bear the brunt of both climate and racial injustice." Cornell, a young black man from Rockaway, also gave a speech – wildly applauded by everyone in the big hall – about social and racial injustices in Rockaway, and then a group of indigenous hip hop artists made a politically charged rap performance. Eventually the whole party moved up to the rooftop, where people danced under the NYC night sky. While the sun slowly rose, I chatted away with activists and residents while making mental notes of everything I would scribble down in my field journal as soon as I got the chance. Observing the weave of trusting connections that spun itself around the rooftop, one would not have thought that conflict between Rockaway residents and activists was hiding beneath the surface. The next section will dig deeper into the disempowerment that took place and that spurred tensions and resistance from some residents.

## Disempowerment Through Misrepresented Subject Positions

The Occupy Sandy activists, deeply aware of structural inequality, aimed to challenge social hierarchies in the group. Race, class, and gender-based inequalities were the main focus. In interviews, many of the Occupy activists shared the view that aggressive capitalism creates persistent socioeconomic inequality, closely intertwined with race-based inequalities. Some of the female Occupy Sandy activists also made a connection between capitalism and gender inequality. Occupy Sandy activist Matthew alleged that since racism is structural, it is therefore impossible to "get out of your racist brain." He also talked about how the hierarchies made their way into the Occupy Sandy hub:

> There's a hierarchy, you can't completely erase the world we live in, there's a hierarchy within our organization. We have different skills, we have different privileges, and we come from different backgrounds.

Borrowing some of the organizing mechanisms from the wider Occupy movement, the activists employed a few compensatory techniques meant to break with inequalities between themselves and the residents. They put demographic restrictions in place regarding who could become a facilitator, with the goal to encourage those identified as being marginalized (people of color, women, and people from low-income communities). In meetings, activists made use of progressive stacks, meaning that they gave the floor first to people from these communities. The activists also engaged in ongoing policing. They called each other out publicly when they thought that a privileged person dominated the conversations or said something that was offensive. Policing was primarily directed toward white male persons in the group. Occupy Sandy activist Aisha was asked to become a facilitator as a result of the demographic restrictions and talked about her mixed feelings about it:

> The other reason for them to bring me on board was that they were a group of mostly white folks who came together to organize black

and Latino people in a community that they didn't know. And it's a little complicated because although it felt a little tokenizing, I also realized that there's value in me coming in doing this process since I'm a person of colour with a working-class background.

Matthew talked about how he, as a white male, tried to sensitize himself to how his own privilege might come into play when he co-facilitated with Aisha:

I can be, like, ok, now I know, I'm going to think about my body language a little bit more, and I'm going to think about the fact that she's got a good bit more experience so I'm going to just play off her a bunch, think about who's talking.

When I talked to the residents about the compensatory mechanisms, some expressed discomfort. It did not always sit well with them when the activists encouraged people of color and/or women to speak first in speaking rounds, or when the activists policed white people and/or men and asked them to step back. Resident Monique said that it had never been a concern to her and expressed the following:

I never much cared about the differences. As much as other people have. I know a couple of people in our team that are very aware of economic difference and racial difference and they're very conscious about, like, in-adverted sexism, they're really aware of it. Way more aware than I have ever planned or intended to be.

Resident Will thought that the compensatory mechanisms triggered a form of distance between the residents and the activists. Well-intentioned as they were, he figured that they hindered equality among members of the group and pinned participants into social categories. The mechanisms only breed "the kind of classism and racism and inter-hatred that was the disaster before the disaster occurred." Some of the residents challenged the Occupy Sandy activists around the issue of compensatory mechanisms and explicitly asked them why they wanted to differentiate themselves by employing different rules for different people in the group. Others expressed a discomfort in interviews but held their tongue in the meetings.

Another aspect that created some confusion for some of the interviewees was the issue of intersectionality. Participants' identities cut across race, class, gender, educational levels, and religious affiliations, and some of these social hierarchies were made explicit through the compensatory mechanisms. These mechanisms, however, were not doing anything to deal with the problem of interlocking processes of power. Intersectional subject positions make compensatory mechanisms difficult to put to practice in a meaningful way. For example, a white woman in the group was considered marginalized based on her gender but privileged based on her race, meaning that a compensatory method was both applicable and non-applicable at the same time.

The most pressing concern that residents brought up was that the compensatory mechanisms did nothing to compensate for the differentiation brought about by the storm itself. In this particular setting – with a storm that had shattered homes, neighborhoods and employment opportunities – race, class, and gender were not the only differentiations at play. Some of the residents were tired, cold, and stressed due to the shattering effects that the hurricane had had on their lives. The situation had caused chronic stress for almost all of the residents. Some had previous substance dependence issues that had flared up; others were either displaced to temporary housing or lived in homes that could not provide basic things such as heat, water, and electricity. On top of that, many were navigating time-consuming bureaucratic hurdles in order to receive the right relief assistance from government and city agencies. Will expressed it as such:

> I don't think they could understand what it is to be able to go home to a nice place, and then they have this here and it's this work that they do. If I'm going anywhere, it's because I'm working on finding relief to the disaster. I haven't taken a vacation from the situation, not once. ... I have not left the situation. I've had no break.

In relation to this, whereas some residents experienced the fluid and meandering meetings as an important way of building community, other residents complained that the meetings were too long. This was especially so for residents who were dealing with getting back on track after the storm and who did not have time nor energy to sit through several hours of meetings. Samantha, a black woman in her late 20s who went to a couple of meetings but then chose to disengage, talked about how the format of the meetings was silly or unstructured. This was frustrating to her since she went to the meetings because she wanted to get something done to rectify what she saw as an unfair life situation for her in relation to the storm.

It seems that the activists' ideas around power and difference went full circle. In trying to accommodate differences in privileges among participants by employing compensatory mechanisms, they made residents uncomfortable by singling them out. The activists let their structural understanding of power and difference guide the way they practically set up the work. However, the different treatment locked participants into social groups, according to some residents. It created unwarranted homogenization by singling them out as representatives of a social group, thus re-creating some of the same tendencies that were the problem to start with. It also made intersectional subject positions invisible. The work was fuelled by a critique of essentialism, yet the critique translated into practice generated its own essentialism.

It seems that focusing too much on social differentiations at micro-level settings may risk placing human beings in locked social identities. Furthermore, the situated marginalization that ensued after the storm had a strong bearing on the issue of influence in the group, yet was not explicitly acknowledged by the activists as a ground for compensatory mechanisms, nor was it taken into account when they planned the format and timeframe of the meetings. This compromised residents' ability to participate on equal terms. The inequalities that are challenged through compensatory mechanisms may not be the only ones at play. If attention is predominantly given to class, race, and gender differentiations, yet situated marginalization is ignored, some participants may still find themselves lacking the capacity to make their voices heard effectively, based on situated marginalization.

## Disempowerment Through Lack of Transparency

The Occupy Sandy activists aimed for an open agenda-setting process, wherein residents would take the lead in coming up with political issues that the group ought to focus on. When I interviewed residents about their experiences of this agenda setting process, some of them voiced harsh appraisal toward the activists. Questions such as "Who decides here?" "Who is the leader?" and "Who signs off on the checks?" were voiced out in my first encounter with the Occupy Sandy network, as described in Chapter 1. Residents such as Jemar expressed concerns about issues they thought were non-transparent, such as the organizational identity of the hub and the financial handling of donations. They also perceived that behind the activists' rhetoric of local ownership was an actual agenda known only to a handful of activists.

### *Organic Organizational Identity and Financial Question Marks*

The process of forming a stable group with a steady presence in Rockaway was an organic and fluid process. Initially, the hub started out as a three-month training project that emerged out of the larger Occupy Sandy network. This was intended to be temporary, and the initiators were supposed to pull out after the training period to leave the work in the hands of the residents. However, the exit from the activists became less straightforward than intended. Some of the activists had built relationships with community members and felt that they wanted to stay on for more long-term work. And so, toward the end of the three-month period, a few of the initiators left while others stayed on to form a long-term group with a steady presence in Rockaway.

The confusion around the stability and identity of the group continued even after this point. There were various perspectives on what kind of long-term group it ought to be across the resident/activist divide and among the activists. The malleable organizational formation created some confusion and distrust among residents. Kiara, a black resident in her mid-30s, was involved in the first three months of the work and then dropped out. She thought that for all the radical discourse that the activists were using, the group was nothing more than one of many non-profit organizations that enter Rockaway to capitalize on the misfortunes of the community, only to leave without contributing to social change. Kiara's notion ties into a broader sentiment of distrust toward outsiders among Rockaway residents – I kept hearing it in interviews, in community board meetings, in informal conversations at local Rockaway events and in reading the letters to the editor of the local newspaper *The Wave*. Resident Jemar thought it was deceptive that the activists had led community members to believe that they were only temporary, whereas in reality they wanted to establish themselves in the area as a long-term group. Some of the activists also raised the confusing organizational status of the hub as a point of contention. Occupy Sandy activist Hannah, a white woman in her late 20s, said that it may have contributed to resident's suspicion and the large number of drop-outs in the first year of the group's existence.

A related issue was the financial status of the hub. A few of the residents conveyed suspicion of how the activists handled financial resources. Most of the Occupy Sandy relief work had been unpaid volunteer work, but the remaining funds from the large amount of money raised during this period was meant to sustain a few of the long-term hubs set up across the city. The funds were primarily used for stipends to the core coordinators in these hubs, among which a majority were Occupy Sandy activists. This allocation of funds was decided through a collective decision-making process in the so-called Occupy Sandy Spokes Council in which representatives from all of the long-term recovery groups from across NYC and New Jersey met regularly. The

Spokes Council model was taken from the OWS movement, where a central organ was supposed to function as a sort of "confederated direct democracy" (Gould-Wartofsky, 2015, p. 124), where each group would send a representative or "spoke" on a rotating basis to deliberate with other spokes. The activists thus received monthly stipends for their work, with an external fiscal sponsor who oversaw the transactions. However, this funding situation was not clearly communicated to the residents.

A local news article also stirred up controversy around the issue of resources, implying that activists had enriched themselves rather than passing fundraised money onto the affected communities (West, 2013). Some of the residents I interviewed were deeply uncomfortable with the stipends. In their view, the activists had raised money for the benefits of the community of Rockaway, yet used most of it to stipend themselves. Others were more concerned with that they did not know where the money was coming from, who was in charge of it, or how it was allocated. The activists were well aware that much of the criticism was for issues of resources. They understood that some of the critical residents saw them as outsiders who were organizing in Rockaway for the money's sake. Yet, activist Matthew stated that the criticism was not legitimate. The stipends were very small, he remarked, and the activists were entitled to them because they had worked hard for free in the relief phase. Resident Chloe was ideologically aligned with Occupy Sandy, yet had chosen to disengage from the hub after the first three months precisely because of this issue. She thought it was unfair to make people believe that they could change the circumstances of their lives without offering them any resources to do so:

> So what you have are people of color who have been excited about the meetings, to think about certain ideas that they may have not thought about or that they may have not felt support around. They may have been thinking about these things already but they didn't have like a bunch of people egging them onto think about it, and applauding them to think about it, and telling them that these are good things and they can happen. There's something wrong with giving people the impression that they have the actual political power to make these changes in their own lives, when they don't. They don't have the opportunities, they don't have the funding, they don't have the support. And then getting funding in their name to then continue to incite them ... I mean essentially, these people are getting paid now, whether they admit it or not, they're being paid through Occupy Sandy as coordinators to stir up people

in Rockaway. And people in Rockaway are getting excited to do things, but they're not getting the money, they're not being funded.

## Hidden Agendas

The lack of transparency in Occupy Sandy's financing was echoed by the flexible agenda-setting. The agenda-setting process was meant to be flexible, thus allowing residents to decide which political issues the group should focus on. Although a few smaller working groups gradually grew out of the initial long-time recovery discussions, the land-use working group received the most attention, and after approximately six months, became the main activity in the hub. Activists and residents investigated city plans for vacant land in Rockaway, made inquiries with land attorneys, and looked into regulations around possibilities for community input in development plans. Bit by bit the other working groups were phased out, and the resources from these extinct working groups were directed toward the land-use group.

After a while, one of the residents took more and more responsibility in the land-use working group. Occupy Sandy activist Elianah, a white woman in her mid-20s, explained how the residents' devotion to the work around land use was a motivation for allocating all of the hub's resources to this group. Their commitment was a symbol of community-led campaign planning, according to Elianah. She explained that once the activists saw that there was someone committed to the work, they fully supported and encouraged them to take the lead. The resident in charge of the land-use group was thus empowered in the process. Besides, the experience changed the residents' views on their own capacities, both in terms of better organizing skills but also in terms of substantial knowledge about urban planning issues.

Some of the other residents were critical of this purportedly flexible agenda setting. They believed that in reality the activists came with an agenda that they pushed through. Julio, a Latino resident of Rockaway in his early 20s, stated the following with regard to this issue:

If you didn't see things in the same way as they did, that was a problem. And people from Rockaway don't see things the same way as some of them and they are pushing certain issues. They have their agenda that is different from what the residents here want to see.

Activist Elianah confirmed this suspicion when she explained that most of the activists had a sense that land use should really be the focus, even if they put up a front of letting the decision be in the hands of the residents:

> I think all of us had a sense that land use should really be the thing. But we were like, "So what do you want to work on?" But I knew it was going to happen. I was thinking, "We'll just say it's working groups and land use will eventually become the thing, the land use will come out of it."

## Disempowerment Through Silenced Resistance

After approximately six months, the hub stabilized into a structure with a core group of core coordinators sustained by stipends. The core group of coordinators was the main decision-making body of the hub. Everyone else were members. Out of the members, there were a few who were considered as emerging leaders that could be added to the core as long as they showed commitment to the group over a period of time. Residents were thus gradually added to the core coordinators group until it had five activists and five residents. Out of the residents who were added to the core group of facilitators, everyone but Will had a positive experience of being encouraged and strengthened. When interviewed, they talked about how their organizing taught them new skills and strengthened them to take on leadership. Monique, for example, explained how she was inspired to take political action in a new way:

> I feel the responsibility to be the change that I want to see in the world. I feel the responsibility to say that if I don't like how things are working, I'm going to change it. And I feel like, I want to do it. Let me change it, let me do it. Let me prove that it can be done! And before, I wasn't really that committed to saying that if I don't like it then let's change it. Before I would say if I don't like it then that's just the way it is. But now I feel like no! I want to change it. Whatever it is.

Will was of another opinion however. He felt that the activists were using what he called "back-door decision-making," where they planned meeting sessions and decided on the agenda behind closed doors, away from the residents. He also raised the fact that the decision to add new people to the core coordinators group lay mainly in the hands of the activists, not the residents. The activists invited residents to the open meetings and continuously looked for certain qualities in them to see who was suitable to be included in the core coordinators group. Some of the interviewees, mainly the dropout residents, believed that the activists were selective in terms of who they encouraged as leaders and were being careful not to add troublemakers to the core group. Samantha wanted to be added to the core group but had to push her way in. She felt that she ought to be part of the core group since "it is my community, my neighborhood," but she was not invited. She described the process as frustrating, feeling as if she was treated differently than other residents:

> The process sometimes felt like in school, where the teachers had their favorite pets among the students and some of the students were just seen as troublemakers – I felt like that's how they saw me, as a troublemaker.

Samantha believed that she was kept away from the core because she was too critical. She later chose to leave entirely. Other residents who had been among the more vocal critics expressed that once they had voiced their concerns, they were being either ignored or subtly silenced by the activists. Kiara, for example, said that since the activists knew each other from before, she avoided voicing her criticism even in private one-on-one conversations because she was afraid that they would talk behind her back and then ice her out for being critical.

In sum, it seems that residents who were non-obstructive – those who did not challenge the activists' views on what the agenda ought to be – were the only ones accepted into the core group of coordinators. The residents who challenged these views were subtly and gradually left outside of the decision-making bodies, and after a while, most of them chose to leave the group altogether. Will, one of the critical residents, stayed on and kept his criticism to himself because he needed the stipends provided. Others left without ever voicing their concerns. A few critical residents engaged in resistance. One strategy was to intentionally sabotage open meetings, as illustrated in the story of my first meeting with Occupy Sandy in the introductory chapter. Apart from sabotaging meetings, critical residents also spread rumors about Occupy Sandy in the wider community, which led to more public displays of disloyalty, large number dropouts among residents, and eventually a crumbling organization.

### *"We're Feeling Pretty Empowered Already"*

Some residents decoded attempts at empowering them as belittling, explaining that they felt patronized by the attempts of the activists to educate them politically and to encourage them to become leaders. Jorge, a black resident in his 30s, with long experience of political activism, puts it as such:

> These mostly white kids put on a documentary about the Black Panthers in an effort to give us a "political education." This was done, mind you, in the middle of a predominantly black neighbourhood... I think that the Occupy organizers erred in this respect, assuming people needed their "training." It was really condescending.

Other residents perceived of the activists as inexperienced in community organizing. Jemar used the term "straight out of college" persons who wanted to gain real-life experience by organizing the communities of Rockaway:

> It's the first time that they're actually doing, things that they've read in books, they went to a seminar last week and they're just acting on these things.

Samantha instead figured that the Occupy Sandy activists had good intentions but expressed a rather dry disbelief in their competencies:

> I think that these folks come in with good intentions, but they have this idea of, 'I've just came out of college and I just finished reading this book, and I'm gonna implement these things, and I'm gonna you know, empower these people!' And then they come and the folks are like, "We're feeling pretty empowered already, you know."

## Conclusion

Occupy Sandy Rockaway was active for about three years until it gradually fell apart, riddled by tension and resistance. The coalition of grassroots organizations also broke down. In the wake of these conflicts, key people, both residents and Occupy Sandy activists left, meetings became scarce, and residents lost interest in the land-use work. Despite Occupy Sandy's awareness of social inequalities and their aim to empower residents, their lack of transparency in organizational and financial issues, and agenda setting demotivated many residents in the recovery period. Moreover, their attempts at correcting social inequalities through compensatory mechanisms were experienced as belittling by some residents. The empowerment that was generally perceived as positive during the relief period was now reserved for core coordinators, and therefore became questioned by Rockaway residents that felt unheard, even unwanted in this post-disaster participatory space.

## References

Adams Sheets, C. (2013). Staten Island's Hurricane Sandy Recovery Slowly Progressing 6 Months Later. *International Business Times*. Retrieved from www.ibtimes.com/staten-islands-hurricane-sandy-recovery-slowly-progressing-6-months-later-1234807

ALIGN [The Alliance for a Greater New York]. (2013, June 12). Bloomberg Storm Plan Praised, But Faces Obstacles. Retrieved from www.alignny.org/posts/clip/2013/06/bloomberg-storm-plan-praised-but-faces-obstacles/

Baxamusa, M. H. (2008). Empowering Communities Through Deliberation: The Model of Community Benefits Agreements. *Journal of Planning Education and Research*, 27, 261–276.

CBA – Community Benefits Agreement. (2015). Draft version of the UPWARD Coalition's Community Benefit Agreement that was circulated internally for editing. Rockaway: Rockaway Wildfire.

deMause, N. (2013, January 31). As Sandy Relief Efforts Fade, Crisis Far From Over [Blog]. Retrieved from http://demause.net/category/environment/hurricanes/

Ellefson, A. (2014, September 12). We want to show what's at stake: Rockaway residents rebuild, demand climate action two years after Superstorm Sandy. Special Issue #200. *The Indypendent* [online newspaper]. Retrieved from www.indypendent.org/2014/09/12/we-want-show-whats-stake-rockaway-residents-demand-climate-action

Enterprise. (2013, October). *Hurricane Sandy: Housing Needs One Year After*. Research Brief. Retrieved from https://s3.amazonaws.com/KSPProd/ERC_Upload/0083708.pdf

Gould-Wartofsky, M. A. (2015). *The Occupiers: The Making of the 99 Percent Movement*. New York: Oxford University Press.

Haygood, B. (2013, October 30). My Birthday Wish. *The Huffington Post*. Retrieved from www.huffingtonpost.com/ben-haygood/my-birthday-wish_1_b_4176009.html

Jaffe, S. (2013, October 29). Whose recovery? A year after Hurricane Sandy hit, despite community efforts, marginalized New Yorkers aren't back on their feet. *In These Times*. Retrieved from http://inthesetimes.com/article/15800/whose_recovery_some_new_yorkers_still_not_back_on_feet_after_sandy

Krauskopf, J., Blum, M., Lee, N., Fortin, J., Sesso, A., & Rosenthal, D. (2013). *Far From Home: Nonprofits Assess Sandy Recovery and Disaster Preparedness*. New York, NY: School of Public Affairs at Baruch College, CUNY Center for Nonprofit Strategy and Management, Baruch College Survey Research, and Human Services Council of New York.

Liboiron, M. (2013). Turning the Tide: Remember Sandy, Revive Our City. Procession and Rally! *Superstorm Research Lab*. Retrieved from http:// superstormresearchlab.org/2013/07/27/turning-the-tide-remember-sandy-revive-our-city-procession-and-rally/

Make the Road New York. (2012). Unmet Needs: Superstorm Sandy and Immigrant Communities in the Metro New York Area. Retrieved from www. maketheroad.org/pix_reports/MRNY_Unmet_Needs_Superstorm_Sandy_ and_Immigrant_Communities_121812_fin.pdf

Murphy, J. (2011, October 29). NYCHA Residents' Unemployment Has Nearly Tripled [Blog post]. Retrieved from www.citylimits.org/blog/blog/171/report-nycha-residents-unemployment-has-nearly-tripled#.UxR9fVMppQU

Rebuild by Design. (2013). Social Vulnerabilities. Retrieved from www.rebuildb ydesign.org/research/vulnerabilities-social

Rohde, D. (2012, October 31). The Hideous Inequality Exposed by Hurricane Sandy. *The Atlantic*. Retrieved from www.theatlantic.com/business/archive/ 2012/10/the-hideous-inequality-exposed-by-hurricane-sandy/264337/

Salkin, P., & Lavine, A. (2008). Negotiating for Social Justice and the Promise of Community Benefits Agreements: Case Studies of Current and Developing Agreements. *Journal of Affordable Housing & Community Development Law,* 17(1/2), 113–144.

SIRR [Special Initiative for Rebuilding and Resiliency]. (2012) [Website] Retrieved from www.nyc.gov/html/sirr/html/home/home.shtml

Solidarity NYC. (2013). Growing a Resilient City: Possibilities for Collaboration in New York City's Solidarity Economy. Retrieved from http://solidarity nyc.org/wp-content/uploads/2013/02/Growing-A-Resilient-City-Solidarity NYC-Report.pdf

West, J. (2013). Occupy Sandy Once Welcomed, Now Questioned. *Mother Jones*. Retrieved from www.motherjones.com/environment/2013/06/occupy-sandy-once-welcomed-now-questioned

# 7  Saviors Trapped in Disaster (Dis-)Empowerment

## A Summary of Sorts

One chilly October morning in 2021, at a coffee house in central Stockholm, I spent a few hours reading and meditating on the concept of empowerment. "Ugh, empowerment, what a shitty concept," Leena Vastapuu, my feminist colleague had quipped the day before when I told her about the book I was writing. "It's just so patronizing." Her remark mirrored my own understanding of empowerment that had grown on me while entangling the narrative of Occupy Sandy in Rockaway. I felt increasingly put off by the concept. Empowerment, such a nice word, but with so many flawed interpretations and outright diluted practical uses. Just then, a white working truck went by on the street. It looked like some sort of electricity company truck. Across its back doors, the company logo was printed in azure-blue letters, along with the marketing slogan: "We Empower Your Day."

Sometimes misused or watered down, other times complicated to put to actual practice; empowerment is at the very least thorny. One dilemma is that empowerment is often initiated and organized by people who do not belong to marginalized communities. This contradicts the empowerment ideal that states that real empowerment emanates from within the community (Campbell, 2014; Cornwall, 2003; McDaniel, 2002; Pilisuk et al., 1996; Snow et al., 2004). This puts some strains on the process of empowerment, as the story of Occupy Sandy in Rockaway shows. Cornwall's notion (2016) of empowerment processes as twisty paths "that can double-back on themselves, meander on winding side-routes and lead to dead-ends as well as opening up new vistas, expanding horizons and extending possibilities" (p. 345) resonates well with the story. In the immediate relief phase, collaboration between Occupy Sandy activists and residents really did expand horizons. The success of the innovative, social justice ingrained relief phase is testimony to

DOI: 10.4324/9781003005278-7

what people can achieve when organized in flexible, autonomous, and horizontal networks. Not only did storm-affected marginalized communities receive shelter, food, warmth, and human connection, they also participated on their own terms, influencing both means and ends of the relief work. Their heightened well-being and increased sense of belonging reflect the results from other researchers who have studied empowerment in disaster relief (Carlton, 2015; Seana & Fothergill, 2009). Active participation shifted residents' roles from receivers of aid to productive partners. Speaking with Dynes (1987), it enabled them to step out of the role as helped and into the role as savior, empowering them along the way.

Looking at the disaster relief period and the question of whether the disaster constituted an opportunity to move away from the problems of inequality – it seems the answer is yes. The special way in which relief was organized contributed to a situation in which marginalized storm-affected communities were able to shake things loose. The problems of inequality were – at least for some time and at least in some aspects – suspended. But if the success of the relief phase opened up new vistas, empowerment in the long-term recovery phase doubled-back on itself. Linking relief to long-term recovery, it seems that the romantic image of participatory and inclusive relief work is perhaps less straight forward than one would think.

To speak with Solnit (2020), Occupy Sandy's transition from relief to recovery was marked by battles, some of which I observed when I met them first time six months after the hurricane. As I witnessed Rockaway residents clamp down on the meeting, I felt an urge to understand their anger and obstruction better. In following the collaboration up-close through participatory observations and interviews, I tried to apprehend what it is like when those who come to empower you are whiter, perhaps richer, younger, and less damaged by the disasters you have seen and felt, both in your daily struggles and in acute predicaments like a devastating storm. In learning from the residents, I found that although they felt empowered through participating in the Occupy Sandy relief work, empowerment doubled-backed on itself when relief shifted into long-term recovery. Yes, a few of the residents felt they learned new skills even in this phase: they became more knowledgeable and got a better handle on organizing. But other residents were harshly critical and deemed the whole process of empowerment completely misguided. They noted how activists operated with pre-set agendas hidden behind a rhetoric of local ownership, felt belittled by the sometimes paternalistic manners of the activists, and experienced that they were subtly silenced, indeed even iced-out when they voiced critique.

## The Savior Trap

I want to propose a concept – the savior trap – to capture how the activists, although deeply aware of their own privileges, nevertheless ended up reproducing some of the same power imbalances they had set out to alter. In thinking about the savior trap, I extend the reasoning around the "(white) savior complex" (Cole, 2015; Randhawa, 2016), which assumes that saviors are unaware of their power position and perceive others as receivers of charity (Davis, 2016). In their quests to save less privileged communities, be it people of color, poor children or women in the Third World, this unawareness leads saviors to reproduce and even exacerbate power imbalances. Contrary to the white savior complex, social justice activists are often acutely aware of this potential problem (Ahmed, 2004; Mahrouse, 2014, Phillips, 1996). As Reinecke (2018) suggests, this awareness propels activists to enact desired futures in the here and now by directly shifting power relations and experimenting with alternative, compensatory mechanisms to disrupt entrenched inequalities. The quest then becomes one of undoing the power of the privileged, and ultimately empowering the marginalized.

Yet, even activists deeply aware of the pitfalls of the savior complex can fall in the savior trap, as illustrated by Occupy Sandy in Rockaway. The story was one in which privileged activists, in the wake of a storm, initiated and organized the processes of empowerment for storm-affected and marginalized communities. Occupy Sandy did not only provide much needed assistance to marginalized communities in Rockaway, but also set in motion a transformative project that aimed to make residents aware of social inequalities so they could take political action. However, despite being cognizant of their own power and working hard to pull it apart, the activists eventually got stuck in their own privileges. By using the concept of the savior trap, I want to convey how activists were eventually unable to break with the inherent power imbued in the relationship between saviors and helped, and therefore unwittingly reproduced their own power. This led to critique, resistance, and an organization that lost its momentum and crumbled.

The savior trap presents deep ethical considerations and puts us between a rock and a hard place. Leaving people to their fate would be far worse; it is not an ethical option. Especially not when disasters breathe down people's necks, reveal stark inequalities and put lives and homes in danger. Yet, when coming to save and empower marginalized communities in the long term, one can get trapped. It can be difficult to accept that you are not the protagonist of the story and that help might not be help at all. The paradox in essence is this: we cannot save, but we cannot *not* save. In the remaining chapter, I weave together three inter-related subsets of the savior trap: the intersectional trap, the resistance trap, and the situated marginalization trap.

Side note. Elaborating on the savior trap, I am using "we," aware that it might read as a way of othering the residents. I am using it because, in a sense, I identify with the activists. As have these activists, I too have learned that society is inherently fucked up. I know I am on the privileged side of things, with my white skin and Swedish name, with my university position, tucked away in the safety of a Swedish middle-class life, with a steady salary coming in every month. The only disaster I have lived through is the Corona Virus Disease 2019 (COVID-19) pandemic and the occasionally flooded basement. All other crises in my life are personal and I have means and resources to deal with them. I can easily relate to the activists' attempts at doing something for the good of others and I can also see the humiliation of trying to put your money where your mouth is, try to actually practice what you preach, yet being shut down by the very same persons you thought you were helping.

## The Intersectional Trap

Back when online activism was new, my friend Viktoria Ask and I ran a feminist collective blog where we invited people to write their own texts about why they had become feminists. Readers and writers of the blog collectively reflected on feminism through personal stories of oppression and awakening. The blog was a success; we got good traffic and some heavy feminist names to come write for us. It felt great to be doing something that resonated with people. Until one day, when we were criticized on Twitter for running a politically obsolete and insular white feminist blog, a critique that was amplified and spread by other feminists of color. It got to me. It got to me hard. Born a human and raised into the gender of girl, living through the soul-crushing sexualization of my early teens, and the internalized objectification and eating disorders that came along with it. Growing up to be a woman, working in male-dominated university settings, always slightly on guard, always fighting to balance the right amount of assertiveness with the risk of being seen as bitchy by male colleagues unaware of their own biases. And then realizing, painfully, that I was on the other side of the equation when it came to racism. I was the biased one, the oppressor, the culprit of exclusion. And I had had no clue about it until I was called out. It was a harsh way to learn about intersecting power dynamics. Some of us are both oppressors and oppressed. Because of it, some of us have a hard time seeing ourselves as oppressors. It is like that optic illusion of a duck that is also a rabbit. It is impossible to see the two at the same time. The intersectional trap is a dilemma that speaks also to the story of Occupy Sandy in Rockaway. On one hand, we cannot look away from how structural power differences may come to bear on who is heard and listened to within social justice spaces (or indeed within any social space). Norms and social codes around how to argue for things are often biased so that only certain people are heard and listened to, whereas others have a hard time generating respect (Mansbridge, 1976). Marginalized people may harbor internalized self-doubt due to experiences of oppression (Kruks, 2001; Meyer, 1995). In trying to offset such power imbalances, feminist scholars and activists alike have suggested that some technique of differentiation is necessary, as discussed in Chapter 3.

On the other hand, as seen when tracing Occupy Sandy in Rockaway, differentiation translated into practice through compensatory mechanisms can generate its own essentialism. Focusing too much on structural differentiations at micro-level settings may risk placing persons (always fluid, often intersectional) in locked social identities. As Phillips (1996) avers, and as Gould (1996) notes, introducing compensatory mechanisms based on structural understandings of difference may risk obscuring pluralism in micro-level settings. Crenshaw (1991), Butler (1990) and Young (2000) discuss this too: social groups are not mutually exclusive. Various social positions intersect in each person. Devising compensatory mechanisms obscures this fact and simplifies identities into locked categories. In the case of Occupy Sandy in Rockaway, participants' identities cut across race, class, gender, educational levels, and religious affiliations. The compensatory mechanisms were thus misguided practically. For if a white woman is under-privileged based on her gender but privileged based on her race – a compensatory method is both applicable and non-applicable at the same time. Perhaps compensatory mechanisms are too crude instruments to be viable in micro-level settings, since in the words of Phillips "diversity is too great to be captured in any categorical list" (Phillips, 1996, p. 146).

### The Resistance Trap

Organic beings (humans, horses, trees, plants) often thrive when having to deal with toxins in moderate doses. This mechanism is called hormesis. Trees never pressured by weather or winds, may topple over and self-die. They need resistance to grow stronger. Heat, cold, radiation, viruses are stressors that, if they do not kill us, make us stronger. Occupy Sandy in Rockaway did however not grow stronger by resistance, but in fact crumbled under the weight of residents' critique and sabotage.

This is in line with Mosse's (2005) ethnographic work illustrating how communities' resistance against British development projects took the form of silent sabotage, and similar to what Campbell (2014) found in a study of the strained relationship between white Western women and black women in the global south. Campbell and Mosse found communities that engaged in subtle manipulation of organizational programming, something that the incoming outsiders could not predict. How are we to understand this form of community agency and resistance? As Mary-Louise Pratt (1991) so eloquently puts it, we need to ask what the place of critique is. Are facilitators of social justice spaces (or classrooms, as Pratt discusses) most successful when they eliminate obstruction and unify "the social world, probably in their own image? Who wins when we do that? Who loses?" (p. 5).

I believe there is a trap lurking here, one of resistance. Resistance in this context may break a social justice space, as it did in Rockaway. Yet, on the other hand, to silence resistance, shut people down and ice them out in order to move along with the saviors' agenda of empowerment, is in fact to throw empowerment out the window. If the only ones empowered are the ones who say yes to pre-set agendas or are silent in their critique – can we even talk about empowerment? Perhaps empowerment is a process that comes with an inherent risk of doubling-back on itself. Social justice spaces geared toward empowerment might very well end up being arenas in which marginalized communities use their agency to resist the very attempt at empowering them. Such resistance may dissolve the whole endeavor. On the other hand, doing nothing means continuing marginalization, unless communities find a way to empower themselves which (as I will discuss in the next section) may be a challenge in light of disasters. We cannot empower without letting critique have its place, yet the critique might cripple the whole empowerment endeavor.

## The Situated Marginalization Trap

Genuine empowerment comes from within. However, to expect marginalized communities to do the heavy work of empowering themselves while simultaneously working around the clock just to survive a disaster is quite the ask. More privileged outsiders need to step in, lend a hand, be of service, or otherwise the disaster may yield nothing but further marginalization. Yet, when stepping in, outsiders risk exacerbating the same power imbalances all over again, because of the disproportionate time, resources and energy they have to influence the process according to their own agendas. Research on social vulnerability has shown us over and again how structural differentiation such as gender, race, age, able-bodiedness, sexual orientation, or identity are particularly salient with regards to disasters (Arora-Jonson, 2011; Enarson & Pearson, 2016; Fothergill & Peek, 2004; Gaillard et al., 2017; Tierney, 2014; Wisner, 2003). In addition to these structural inequalities, there might be situational imbalances at play which bear on whose voice is heard. Kruks (2001) focuses on the lived experiences of bodily pain and fear that can come to constitute cognition, judgment, and speech. Although she hones in on bodily experiences of motherhood and domestic violence, I want to add that lived experiences of disasters could be seen through a similar lens. Being exposed to a storm that shatters your home, destroys your work opportunities, or separates you from your nearest is a lived experience of pain that does a number on your capacity to act or speak.

To sum it up, the savior trap goes: we cannot save, but we cannot *not* save. The intersectional trap specifies this further. On one hand, we cannot save marginalized people by acknowledging differences. On the other hand, we cannot *not* acknowledge differences, because they matter. The trap of resistance is that it can break an empowerment process into pieces. However, when resistance is silenced and people are shut down, this also puts a stop to empowerment. In a sense, the baby is easily thrown out with the bath water. The situated marginalization trap is this: marginalized people exposed to disasters are in need of outside aid that empower them to shift their roles from helped to helpers. Yet, given both the structural and situational power imbalances that permeate post-disaster processes, such help may risk exacerbating the imbalance rather than offset it.

## A Trapped Disaster Scholar

The story of Occupy in Rockaway, as seen through the eyes of both activists and residents, has a lot to tell about how well-meaning social justice activists, often popping up in the aftermath of disasters, can turn the altruism and sense of community that often mark disaster relief periods into something sour – despite the best of intentions. When relief shifts into long-term recovery, and things are getting back to normal, and as emerging groups turn into more established organizations, power imbalances, temporarily suspended as people unite in the face of a common challenge of creating safety, may find their way back into the collaborative processes.

In a sense, this is nothing new. The emancipatory literature has raised similar problems before. Some social justice spaces have been shown to be nothing more than arenas in which privileged people dominate the agenda, meaning that they reproduce the same problems of inequality that were the issue to begin with (Cornwall, 2003). What we learn in exploring Occupy Sandy in Rockaway is that post-disaster processes may alter these tendencies. Because of the urgency and share magnitude of the practical problems that come along in the wake of disasters, it seems that breaching inequality may actually be possible, at least in a short time span. However, as things go back to normal, so do social relations, with all of their hierarchies and biases. And along with the return to normalcy, the saviors that set out to undo their own power are boomeranged back into the confines of their privileges.

This books taps into a discussion about the micro-level dynamics and challenges involved in breaching inequality in light of disasters. These are issues of high relevance as we move toward harsher climate related disasters at the same time as social, economic, and political inequalities widen. Disaster contexts may become the new normal. Differentiated vulnerability to disasters is increasingly coming to the fore of social movements aiming to empower the powerless. Mobilization around climate change is happening as communities are experiencing the effects of climate change through more frequent, less predictable, and sometimes harsher weather and climate-related disasters (Aalst, 2006; Field et al., 2012; Ripple et al., 2019).

In this book, I dived deep into a single social justice space triggered by Hurricane Sandy. I entangled the complexities of the social relations and hierarchies within this space while also tying it to the particular place where the process played out: Rockaway in Queens, NYC. But even if the empowerment process would have succeeded, it would be quite the ask to expect such a small-scale pocket of social justice to make largely more than a dent in the structural and historical forces that produce social vulnerability to disasters. Genuine transformation would mean solutions that span the spectrum from global to local. It would, for example, require more regulated markets so that the extreme economic growth paradigm within our capitalist societies can be reined in; greenhouse gas emissions would need a serious halt; and a thorough redistribution of wealth from privileged to less privileged groups would need to be enforced. A transformation would therefore require that the international community overcome the collective action problem that is climate change mitigation and adaptation, as well as substantially different urban planning in many countries so to get rid of discriminatory effects of housing and dwelling that result in unequal exposure to disasters. On top of that, a deep transformation needs a complete shutdown of the racism, ethnocentrism, sexism, and ableism that is part and parcel of many societies. In brief, transformation calls for a complete reconfiguration of the social system as we know it. Even so, I would argue that micro and macro levels make each other interesting. If we only train our eyes on the macro, applying nothing but an eagle's perspective, transformation may seem unattainable in light of the historical and structural forces at play. Yet, the micro-level processes are interesting for their potential for empowerment. In them, we might either find the seeds for change or encounter dynamics – such as savior traps – that hinder political organizing toward macro-political transformation.

This book could be read as a critique of do-gooders. This is far from my intention. I have explored a practically and theoretically interesting problem – that there may be something inherently contentious about trying to empower someone else. I ended up spending years combing the insides of a winding, organically shifting, and sometimes very messy process in which this problem was at the core. People who chose to engage in empowerment projects within social justice movements are often driven by a genuine frustration with being on the privileged side of the equation, and they most often honestly wish to make a difference. The activists I learned from devoted months, even years, working toward something they believed would make a difference. Turned out they could not avoid the savior trap. But at least they tried.

In writing these conclusions, I feel somewhat trapped myself. I have explored what could be said to be a most likely case for post-disaster empowerment. I encountered social justice activists deeply aware of the pitfalls of being a savior, who worked adamantly to undo their own power in the wake of a storm that had struck unevenly across their city. Yet they still got stuck in a savior trap. I am struggling with the implications of my study. Because what is the alternative? Certainly not a call for more hierarchical and bureaucratic organizations – I think it is safe to say those kinds of organizations do very little to empower disaster-affected and marginalized communities. Perhaps there is no alternative other than to keep trying. To try to constantly listen and be willing to change in light of the criticism, skepticism, and resistance that may come from the ones we try to save. Perhaps it is in that resistance that the key to real empowerment lies. Perhaps we cannot save, but we cannot stop trying.

# References

Aalst, M. K. van (2006). The impacts of climate change on the risk of natural disasters. *Disasters*, 30(1), 5–18.

Ahmed, S. (2004). Declarations of whiteness: The non-performativity of anti-racism. *Borderlands*, 3(2). Retrieved from www.borderlands.net.au/vol3no2_2004/ahmed_declarations.htm

Arora-Jonson, S. (2011). Virtue and vulnerability: Discourses on women, gender and climate change. *Global Environmental Change*, 21, 744–751.

Butler, J. (1990). *Gender Trouble*. New York: Routledge.

Campbell, C. (2014). Community mobilization in the 21st century: Updating our theory of social change? *Journal of Health Psychology*, 19(1), 46–59.

Carlton, S. (2015). Connecting, belonging: Volunteering, wellbeing and leadership among refugee youth. *International Journal of Disaster Risk Reduction*, 13, 342–349.

Cole, T. (2012, March 21). The White-Savior Industrial Complex. *The Atlantic*. Retrieved from www.theatlantic.com/international/archive/2012/03/the-white-savior-industrial-complex/254843/

Cornwall, A. (2003). Whose voices? Whose choices? Reflections on Gender and Participatory Development. *World Development*, 31(8), 1325–1342.

Cornwall, A. (2016). Women's empowerment: What works? *International Development*, 28(3), 342–359.

Crenshaw, K. (1991). Mapping the margins: Intersectionality, identity politics, and violence against women of color. *Stanford Law Review*, 43(6), 1241–1299.

Davis, A. (2016). *Freedom Is a Constant Struggle: Ferguson, Palestine and the Foundations of a Movement*. Chicago: Haymarket Books.

Dynes, R. R. (1987). The concept of role in disasters. In R. R. Dynes, B. de Marchi, & C. Pelanda (Eds.), *Sociology of Disasters: Contribution of Sociology to Disaster Research* (pp. 71–103). Milano, Italy: Franco Angeli.

Enarson, E., & Pearson, B. (Eds.) (2016). *Men, Masculinities and Disasters*. New York: Routledge.

Field, C. B., Barros, V., Stocker, T. F., Qin, D., Dokken, D. J., Ebi, K. L., ... Midgley, P. M. (Eds.) (2012). *Managing the Risks of Extreme Events and Disasters to Advance Climate Change Adaptation. A Special Report of Working Groups I and II of the Intergovernmental Panel on Climate Change*. Cambridge, UK: Cambridge University Press.

Fothergill, A., & Peek, L. (2004). Poverty and disasters in the United States: A review of recent sociological findings. *Natural Hazards*, 32, 89–110.

Gaillard, J. C., Sanz, K., Balgos, B. C., Dalisay, S. N. M., Gorman-Murray, A., Smith, F., & Toelupe, A. (2017). Beyond men and women: A critical perspective on gender and disaster. *Disasters*, 14(3), 429–447.

Gould, C. C. (1996). Diversity and democracy: Representing differences. In S. Benhabib (Ed.), *Democracy and Difference: Contesting the Boundaries of the Political* (pp. 171–187). Princeton: Princeton University Press.

Kruks, S. (2001). *Retrieving Experience: Subjectivity and Recognition in Feminist Politics*. Ithaca: Cornell University Press.

Mahrouse, G. (2014). *Conflicted Commitments: Race, Privilege, and Power in Solidarity Activism*. Montreal: McGill Queen's University Press.

Mansbridge, J. (1976). Conflict in a New England town meeting. *The Massachusetts Review*, 17(4), 631–663.

McDaniel, J. (2002). Confronting the structure of international development: Political agency and the Chiquitanos of Bolivia. *Human Ecology*, 30(3), 369–396.

Meyer, I. (1995). Minority stress and mental health in gay men. *Journal of Health and Social Behavior*, 36(1), 38–56.

Mosse, D. (2005). *Cultivating Development: An Ethnography of Aid Policy and Practice* (Anthropology, Culture and Society Series). Ann Arbor, MI: Pluto Press.

Phillips, A. (1996). Dealing with difference: A politics of ideas or a politics of presence? In S. Benhabib (Ed.), *Democracy and Difference: Contesting the Boundaries of the Political* (pp. 139–153). Princeton: Princeton University Press.

Pilisuk, M., McAllister, J., & Rothman, J. (1996). Coming together for action: The challenge of contemporary grassroots community organizing. *Journal of Social Issues*, 52(1), 15–37.

Pratt, M. L. (1991). Arts of the contact zone. *Profession, 91*, 33–40.

Randhawa, S. (2016). Poverty porn vs empowerment: The best and worst aid videos of 2016. *The Guardian*. Retrieved from www.theguardian.com/glo bal-development-professionals-network/2016/dec/08/radiator-award-pove rty-porn-vs-empowerment-the-best-and-worst-aid-videos-of-2016

Reinecke, J. (2018). Social movements and prefigurative organizing: Confronting entrenched inequalities in Occupy London. *Organization Studies*, 39(9), 1299–1321.

Ripple, W. J., Wolf, C., Newsome, T. M., Barnard, P., & Moomaw. W. R. (2019). World scientists' warning of a climate emergency. *BioScience*, 70(1), 8–12.

Seana, S., & Fothergill, A. (2009). 9/11 volunteerism: A pathway to personal healing and community engagement. *Social Science Journal*, 46, 29–46.

Snow, D. A., Soule, S. A., & Kriesi, H. (Eds.). (2004). *The Blackwell Companion to Social Movements*. Hoboken, NJ: John Wiley & Sons.

Solnit, R. (2020, March 31). What Disasters Reveal About Hope and Humanity [interview by H. McQuilkin and M. Chakrabarti]. *WBUR On Point*. Retrieved from www.wbur.org/onpoint/2020/03/31/rebecca-solnit-hope-coro navirus

Tierney, K. (2014). *The Social Roots of Risk: Producing Disasters, Promoting Resilience*. Stanford: Stanford University Press.

Wisner, B. (2003). Sustainable suffering? Reflections on development and disaster vulnerability in the post-Johannesburg world. *Regional Development Dialogue*, 24(1), 135–148.

Young, I. M. (2000). *Inclusion and Democracy*. Oxford: Oxford University Press.

# Index